REORGANIZING HEALTH CARE DELIVERY SYSTEMS: PROBLEMS OF MANAGED CARE AND OTHER MODELS OF HEALTH CARE DELIVERY

RESEARCH IN THE SOCIOLOGY OF HEALTH CARE

Series Editor: Jennie Jacobs Kronenfeld

Recently published volumes:

Volume 16: Quality, Planning of Services, and Access Concerns: Impacts on Providers of Care, Health Care Institutions, and Patients, 1999

Volume 17: Health Care Providers, Institutions, and Patients: Changing Patterns of Care Provision and Care Delivery, 2000

Volume 18: Health, Illness, and Use of Care: The Impact of Social Factors, 2000

Volume 19: Changing Consumers and Changing Technology in Health Care and Health Care Delivery, 2001

Volume 20: Social Inequalities, Health and Health Care Delivery, 2002

RESEARCH IN THE SOCIOLOGY OF HEALTH CARE
VOLUME 21

REORGANIZING HEALTH CARE DELIVERY SYSTEMS: PROBLEMS OF MANAGED CARE AND OTHER MODELS OF HEALTH CARE DELIVERY

EDITED BY

JENNIE JACOBS KRONENFELD

Department of Sociology, Arizona State University, USA

2003

ELSEVIER
JAI

Amsterdam – Boston – Heidelberg – London – New York – Oxford – Paris
San Diego – San Francisco – Singapore – Sydney – Tokyo

ELSEVIER Ltd
The Boulevard, Langford Lane
Kidlington, Oxford OX5 1GB, UK

First edition 2003

Library of Congress Cataloging in Publication Data
A catalogue record from the British Library has been applied for.

ISBN: 0-7623-1069-3
ISSN: 0275-4959 (Series)

⊗ The paper used in this publication meets the requirements of ANSI/NISO Z39.48-1992 (Permanence of Paper).
Printed in The Netherlands.

CONTENTS

PART II: SPECIAL GROUPS OF PATIENTS AND HEALTH ISSUES

PART III: LESSONS FROM OTHER COUNTRIES

PART IV: BROADER POLICY CONCERNS AND HEALTH INSURANCE REFORM

LIST OF CONTRIBUTORS

Denise Anthony	Department of Sociology, Dartmouth College, Hanover, USA
Jane Banaszak-Holl	Department of Health Management and Policy School of Public Health, University of Michigan, USA
Zachary W. Brewster	Department of Sociology and Anthropology, North Carolina State University, USA
Donna J. Brogan	Biostatistics Department Rollins School of Public Health, Emory University, Atlanta, USA
Elizabeth Furlong	School of Nursing, Center for Health Policy and Ethics, Creighton University, Omaha, USA
James W. Grimm	Department of Sociology, Western Kentucky University, Kentucky, USA
Brenda L. Haile	College of Nursing, Texas Woman's University, Houston, USA
Joseph A. Kotarba	Department of Sociology, University of Houston, Texas, USA
Nancy G. Kutner	Department of Rehabilitation Medicine Emory University, Atlanta, USA
Peggy Landrum	College of Nursing, Texas Woman's University, Houston, USA

Richard Lockwood	Department of Sociology, Portland State University, Portland, USA
Blossom Y. J. Lin	Department of Health Administration, China Medical College, Taichung, Taiwan, ROC
Karen Lutfey	Robert Wood Johnson Foundation Scholars in Health Policy, Research Program School of Public Health Berkeley, USA
Teresa L. Scheid	Department of Sociology, Anthropology and Social Work, The University of North Carolina at Charlotte, USA
Karen Seccombe	Center for Public Health Studies, School of Community Health Portland State University Portland, USA
Katherine Clegg Smith	Research Specialist, Health Research and Policy Centers, University of Illinois at Chicago, USA
D. Clayton Smith	Department of Sociology, Western Kentucky University, Kentucky, USA
Debra Trimble	College of Nursing, Texas Woman's University, Houston, USA
Thomas T. H. Wan	Doctoral Program in Public Affairs and Health Services Administration, College of Health and Public Affairs, University of Central Florida, Orlando, USA
Diane L. Zablotsky	Department of Sociology, University of North Carolina at Charlotte, USA

INTRODUCTION

The theme of this volume is reorganizing health care delivery systems: problems of managed care and other models of health care delivery. The volume contains 11 papers, organized into four sections. The sections cover, in order, managed care issues and organizational features, special groups of patients and health issues, lessons from other countries, and broader policy concerns and health insurance reform. Issues of how to best organize a health care delivery system are not new, but the amount of interest in this topic in the U.S. (as well as in other countries) has grown in recent decades. Reorganizing health care delivery systems is a concern of many of the systems of the world, and this volume contains some papers from countries other than the U.S., although the majority of the papers do relate issues to the U.S. health care delivery system. While the majority of the papers in this volume relate to structural and organizational factors, the impact of individual patients is not neglected. One section focuses very much on more specialized groups of patients, and several other papers use surveys of patients or other individual level data as part of their discussion of these issues. Before discussing the specific papers briefly, it is useful to review some material about the organization of the health care delivery system, especially within the U.S. This includes some discussion of how issues of organization of care and special forms of delivering care such as managed care have become major issues in the U.S. health care system at this point in time.

One of the continuing issues in the sociology of health care is how to better organize and reorganize systems of health care delivery. Issues and concerns about how best to organize the delivery of care are not limited only to the U.S., although the majority of the papers in this volume relate to the U.S. health care system. But concerns about having the best system for a reasonable amount of money and assuring that the quality of the care delivered is high are concerns of most health care systems of the world today. Within the U.S., much of the issue has become linked with managed care and with various models of how to reorganize health care delivery systems. A focus on managed care has been part of reforms of the health care delivery system that have emphasized structural reform (Scott, Mannion, Davies & Marshall, 2003). In other countries, such as Great Britain, an example of reforms focused on structural change was the establishment of standard setting bodies such as the National Institute for Clinical Excellence

(Department of Health, 1998). More recently, some analysts have argued that structural changes alone do not always deliver anticipated improvements in quality and performance of health care, and changes in organizational culture may also be required (Shortell, Bennett & Byck, 1998; Scott, Mannion, Davies & Marshall, 2003). Papers in this volume look at structural factors, but some also continue other aspects of organizations and how they function and patient issues as well.

In the past, one of the negative comments that has been discussed concerning delivery of health care in the U.S. has been the lack of a clear system of care. Critics have argued that, within the U.S., there have been different models of how people find health care depending on whether the people had health insurance, the type of health insurance they had, and whether the payer for health care was a government program such as Medicare for the elderly and disabled or Medicaid, for those of limited economic means. In many ways, this is still true today, but there is more attention to explicit models of health care delivery than there was 35 years ago, for example, which was around the time that the Medicare and Medicaid programs in the U.S. were first enacted. Those programs represented the first major government role in the direct provision of care in the U.S. This new role for government provided a contrast with many European countries. In those countries, the role of government in the direct provision of care had already existed for decades, sometimes through a total government organized and financed system, such as in Great Britain, and sometimes through a complicated set of quasi-governmental sickness funds and government programs in Germany and France.

Today, the U.S. is generally viewed as having the most massive, complex and costly health care services system in the world. The debate about whether the U.S. really has a system of care or is a system in a true sense has been discussed now for decades (Shi & Singh, 2001; Wolinsky, 1988) and is not truly resolved. If one compares the U.S. to most developed countries in which there is some type of guarantee of health insurance, most often with programs run by the government and financed at least partially through general taxes, then the U.S. does not really have a system or at least not one system of care. How employed people receive services varies greatly, and those services are generally paid for by some type of insurance company with premiums funded partially by the employee and partially by the employer. For those who receive services through government programs such as Medicare, Medicaid and CHIP, the funding base varies. For Medicare, there are taxes linked to this program that are withheld from each person's paycheck. In addition, current recipients of Medicare pay a premium that is generally deducted from their Social Security (government retirement) check each month. For Medicaid recipients, much of the funding is from current tax revenues, with the responsibility for those revenues split between the national (federal) government and the state governments. For CHIP, a similar model of funding is in place, but some states also charge premiums for some of the children eligible for the program.

One of the biggest problems with the health care delivery system in the U.S. is that not all people have health insurance that provides good access to care. In the past few years in the U.S., it has become clear that the proportion of Americans without health insurance at some point has been increasing. Most experts agree that the reasons for the rising numbers of uninsured are the sluggish economy and rising health care costs. Recent estimates are that 75 million people, or 30.1% of the population, were without health insurance at some point during 2001 or 2002 (Going Without Health Insurance, 2003; Meckler, 2003). The declining economy means that rates of uninsurance are increasing now, after having dropped from 1998 through 2000. However, if one deletes the elderly population, the vast majority of whom are covered by Medicare, the percentage of people without health insurance coverage at some point increases to over a third of people younger than 65. Of those people without health insurance, 90% are uninsured for at least three months and 80% were in working families (Going Without Health Insurance, 2003). A variety of studies have found that people without health insurance are less likely to see doctors and more likely to be diagnosed with illnesses late.

Because the number of people without health insurance and therefore without good access to health care has again been growing in the U.S., problems with quality of care and actual services received have also become issues. Although the U.S. spends more on health care than any other industrialized country, Americans do not necessarily receive more care. Yet the amount spent is much higher. The country with the second highest per capita expenses is Switzerland, and the U.S. spends 44% more than that country (United States Spends Substantially More On Health Care Than Any Other Country, 2003). The U.S. spends 83% more than its nearest neighbor, Canada, and 134% higher than the median of 1,983 dollars for the 30 industrialized countries in the Organization for Economic Cooperation and Development (OCED). The amount of money financed from public sources, such as Medicare, Medicaid, and CHIP, is actually similar to other countries. In the U.S., public sources of funding accounted for 5.8% of GDP in 2000 as compared with 5.9% in United Kingdom, Italy and Japan and not much smaller than the Canadian figure of 6.5% (Anderson, Reinhardt, Hussey & Petrosyan, 2003). The difference in spending is mostly caused by higher prices for these goods and services in the U.S., according to the authors of this study.

How do these facts and figures link with managed care? Despite the growth in managed care in the U.S. in the 1990s as a way to control health spending, the spending gap between the U.S. and other industrialized countries actually widened slightly between 1990 and 2000 (Anderson et al., 2003). Managed care can best be thought of as an effort by employers, the insurance industry and some of the medical profession to establish priorities about use of the health care system and decide who gets what from health care. In an article discussing the end of managed care, Robinson (2001, p. 2622) argues that "after a turbulent decade of trial and error, that experiment can best be characterized as an economic success but a political failure." One fundamental question is how will difficult decisions be made about priorities in health care spending, given an assumption that the nation lacks enough economic resources to finance all services that might provide any benefit to some patient. Managed care in the decade of the 1990s did help to reduce prices, at least temporarily, and reduced the use of some unnecessary services enough to flatten health care expenditures for a while. Part of this success, however, may have been linked to general economic trends in which general inflation became very low, thus also holding down health care cost inflation. Moreover, some of these economic successes have occurred along with rising consumer dissatisfaction about restrictions. While consumers welcome comprehensive benefits and better coverage of medications, for example, they resent many of the rules and regulations. These include such things as utilization review processes that make them wait to find out if certain procedures will be authorized and formularies for drugs that dictate whether a prescription written by a physician will be covered or must be rewritten for a drug covered by the managed care company. As a reaction to these complaints by both consumers and employers and providers, many managed care contracts now have broadened physician panels, removed some restrictions and even in some cases modified the payment structure to physicians. Robinson argues that the consumer is emerging as the locus of priority setting in health care. He believes this shift to consumerism is driven by a widespread skepticism of governmental, corporate and professional dominance.

A different review of recent changes in managed care explored how the organization and dynamics of health systems changed between 1999 and 2001, given the context of expectations from the mid-1990s when managed care was in the ascendance (Lesser, Ginsberg & Devers, 2003). The authors argue that the unprecedented, sustained economic growth of the mid-1990s has resulted in tight labor markets, making employers highly responsive to consumer demands for fewer restrictions on access to care. As part of this, health plans have moved away from core strategies and providers have gained some leverage relative to managed care plans. As contrasted with Robinson's emphasis on consumers, this analysis looks more at the changes for providers, but also sees that the goals of cost containment and quality improvement may be more difficult to meet in the future.

The health care system is a dynamic system, and, just as analysts begin to review the experiences of new approaches, some factors begin to change. Since 2001, cost pressures have started to reoccur, but choice appears to still be valued. Health insurers now are responding to these new cost pressures by giving enrollees choices (Robinson, Yegian, Ginsberg & Priselac, 2003). Hospital tiering is occurring, along with pharmaceutical tiering. In pharmaceutical tiering, patients pay a lower co-payment for generic drugs and higher ones for more expensive name brand drugs and newer drugs. For hospitals, some are designated "core" or premium providers, based on a variety of factors such as costs, quality and structural factors. Enrollees in preferred provider organizations may pay a higher coinsurance rate for using a hospital in the premium tier rather than one in the core tier. Tiered networks make managed care even more complex for consumers.

Many of these recent papers have not paid much attention to the more sociological types of factors that relate to managed care. These factors include things such as how people's relationships and networks impact economic behavior in seeking out health care as well as how power is organized and how state and political values frame markets and competitions (Light, 2000). In this book, the articles will cover some of these organizational issues and patient-related issues as well as lessons from other countries and broader policy concerns. In addition, one section focuses on more specialized groups of patients.

Part one focuses on managed care and organizational features and includes four papers on these topics. The first paper by Brogan reports on a survey of members of health insurance plans and discusses a variety of important methodological issues as well as some substantive ones. Most importantly, the sampling approaches used by contractors for health care plans often result in biased estimates of population parameters, thereby reducing the utility of the information. The paper includes suggestions for improving the quality of information obtained. Anthony and Banaszak-Hall examine organizational variations in the managed care industry during the 1990s through the use of national censuses of managed care organizations at two different points in the 1990s. They draw upon institutional theory to develop a framework to better understand why certain organizational forms and practices emerged at the various timepoints. They find that over time there is increasing diversity in the organizational forms and practices of managed care. Grimm, Brewster and Smith bring a more theoretical approach to the examination of managed care experiences, education and health outcomes. Drawing upon the "sense of coherence" approach, they provide a better understanding of the multiple effects of education on health and how managed care experiences complicate that relationship. In the final paper in this section, Lutfey uses ethnographic data to examine how organization features of medical settings are connected to both the cognitive and interactional work of medical providers. While using a different methodological approach than the other three papers, this paper focuses on an important aspect of care – patient adherence and its relationship to treatment decisions.

Part 2 of the book includes three papers that each have a focus on some special group of patients and health care issues. All three of the papers deal with aspects of reorganizing health care delivery systems, and two of them also relate to managed care issues. One paper examines issues of managed care and HIV care for women, one focuses on management of chronic diseases, and on focuses on the challenges of a changing environment in health care for an important sector of care, long term care. The paper by Kotarba, Haile, Landrum and Trimble contributes to the understanding of women's experiences of living and surviving with AIDs. The paper integrates concepts from nursing and medical sociology, specifically the idea of inner strength as a way people deal with illness experiences with the sociological concept of the existential self. Kutner focuses on patients dependent on a life-maintaining technology such as kidney dialysis. She demonstrates how providers tend to use a cure-oriented model that allows the providers to dominate the care system. By incorporating health promotion and rehabilitation models, patients may be empowered to collaborate with clinicians to maximize functioning and patient well-being. The last paper in this section by Scheid and Zablotsky looks at the ways that managed care tries to limit access to costly services such as long term care. They collected data from long term care agencies in an urban and suburban county to determine the degree of managed care penetration and administrators' evaluation of managed care.

Part 3 includes two papers focusing on the health care systems in other countries and those models of health care delivery. Because these papers deal with countries other than the U.S., they do not discuss managed care in the American context explicitly. One paper focuses on Great Britain and the other on Kazakhstan. Smith conducted 18 months of ethnographic fieldwork to explore the role of general practitioners in the implementation of a major health care reform in Great Britain, the New NHS. She analyzed local implementation processes in relationship to sociological understanding of the role of professions in reform. Contrary to what is predicted by sociological theory, she found that local professionals and health care managers engaged with policy objectives and worked to protect the status quo but within the boundaries of the official reform rhetoric. Wan and Lin used a social capital approach to examine the determinants of health services use as a way to improve planning for effective services in the newly independent state of Kazakhstan. They find that health status is a strong predictor of health services use when the effect of social capital is held constant, and that social capital is directly linked with health status.

The last section of the book includes two papers with the broadest policy focus. Furlong describes the changing political milieu in the U.S. and explores what factors might push the U.S. towards universal health care and what such a plan might be. The article presents evidence-based outcomes of multiple measures of concern with the U.S. health care system, focusing on the issues of access, quality and cost. Seccombe and Lockwood discuss issues of life after welfare in rural communities and small towns and relate this to health insurance concerns. In the last decade, one major social policy change in the U.S. has been the major change in welfare with the creation of the Temporary Assistance to Needy Families Program (TANF) and its concomitant expiration of benefits including health insurance (Medicaid). Through focus groups, respondents reported that topics related to health insurance and planning for health insurance were not covered as part of welfare to work curriculum nor were these issues discussed by caseworkers. As people begin to leave TANF and Medicaid, there may be major concerns about growth in uninsurance rates in this group of lower income workers. Until the more major reforms such as Furlong discusses are implemented, policy analysts and sociologists need to be concerned about access to care for this vulnerable population group.

All the papers in this volume deal with important current issues in medical sociology, with a focus on reorganizing health care delivery systems. Some have explicit policy points of view, such as the two papers in the fourth section. Others relate to narrower, but still important, issues in health care delivery and provide a sociological perspective on those concerns such as those in the second part of the volume. Two papers provide a sociological perspective on issues in the organization and delivery of health care in countries other than the U.S. and provide lessons for the U.S., just as the U.S. experience provides lessons for other countries. The papers in the first portion of the volume focus more on managed care approaches, one of the current solutions in the U.S. receiving the most attention. One of the lessons of the past decade is that no countries in the world are completely satisfied with their health care delivery systems nor do they believe they have resolved some of the tensions of how to provide access to all to high quality care at a reasonable cost. In different ways, each of the papers in this volume contribute to improving our understanding of these issues of health care delivery and provide a sociological focus for the exploration of these issues.

Jennie Jacobs Kronenfeld
Series Editor

REFERENCES

Anderson, G. F., Reinhardt, U. E., Hussey, P. S., & Petrosyan, V. (2003). It's the prices stupid: Why the United States is so different from other countries. *Health Affairs*, *22*, 89–105.

Department of Health (1998). *A first class service, quality in the NHS*. London: Department of Health.

Going Without Health Insurance (2003). www.familiesusa.org/site/docserver/Going_Without_HealthInsurance.pdf

Lesser, C. S., Ginsberg, P. B., & Devers, K. J. (2003). The end of an era: What became of the managed care revolution in 2001? *Health Services Research*, *38*(Part II), 337–355.

Light, D. W. (2000). The sociological character of health-care markets. In: A. Gary, R. Fitzpatrick & S. C. Scrimshaw (Eds), *The Handbook of Social Studies in Health and Medicine*. Thousand Oaks, CA: Sage.

Meckler, L. (2003). Thirty percent in U.S. without health care coverage in 2002–2003. *Arizona Republic*. March 5th, A17.

Robinson, J. C. (2001). The end of managed care. *JAMA*, *285* (May 23rd–30th), 2622–2628.

Robinson, J. C., Yegian, J. M., Ginsberg, M. E., & Priselac, T. M. (2003). Tiered hospital networks in health insurance: Experiments in cost control. *Health Affairs*, *22*, 9–10 and www.healthaffairs.org/WebExclusives/CHCF_Web_Excl_031903.htm

Scott, T., Russell, M., Davies, H., & Marshall, M. (2003). The quantitative measurement of organizational culture in health care: A review of the available instruments. *Health Services Research*, *38*, 923–945.

Shi, L., & Singh, D. A. (2001). *Delivering health care in America: A systems approach*. Gaithersburg, MD: Aspen Publishers.

Shortell, S. M., Bennet, C. L., & Byck, G. R. (1998). Assessing the impact of continuous quality improvement in clinical practice: What will it take to accelerate progress? *Milbank Quarterly*, *76*, 593–624.

United States Spends Substantially More On Health Care Than Any Other Country (2003). www.healthaffiars.org/press/mayjune0301.htm

Wolinsky, F. D. (1988). *The sociology of health: principles, practitioners and issues* (2nd ed.). Belmont, CA: Wadsworth Publishing Company.

PART I: MANAGED CARE ISSUES AND ORGANIZATIONAL FEATURES

SURVEYS OF MEMBERS OF HEALTH CARE INSURANCE PLANS: METHODOLOGICAL ISSUES

Donna J. Brogan

ABSTRACT

Health care insurance companies often conduct sample surveys of health plan members. Survey purposes include: consumer satisfaction with the plan and members' health status, functional status, health literacy and/or health services utilization outside of the plan. Vendors or contractors typically conduct these surveys for insurers. Survey results may be used for plans' accreditation, evaluation, quality improvement and/or marketing. This article describes typical sampling plans and data analysis strategies used in these surveys, showing how these methods may result in biased estimators of population parameters (e.g. percentage of plan members who are satisfied). Practical suggestions are given to improve these surveys: alternate sampling plans, increasing the response rate, component calculation for the survey response rate, weighted analyses, and adjustments for unit non-response. Since policy, regulation, accreditation, management and marketing decisions are based, in part, on results from these member surveys, these important and numerous surveys need to be of higher quality.

Reorganizing Health Care Delivery Systems: Problems of Managed Care and Other Models of Health Care Delivery
Research in the Sociology of Health Care, Volume 21, 3–19

ISSN: 0275-4959/doi:10.1016/S0275-4959(03)21001-2

INTRODUCTION

Employers, governments and consumers use evaluative data on health care insurance plans to make purchase decisions and to decide whether to renew their plan choice. Health care insurers use evaluative data on their health plans for accreditation, quality improvement, marketing, sales, and comparison to other plans. Components of this evaluative data are based on sample surveys of plan members. Common purposes of these sample surveys are to assess: (1) member satisfaction with the plan; (2) functional/health status or health literacy of members; (3) plan disenrollees' reasons for leaving the plan; and (4) members' potential use of specific health education or clinical services.

This paper describes the typical survey methodology recommended for members of health plans. Limitations and problems of this methodology, as implemented, are identified in three specific areas: sampling techniques, field operations and data analytic strategies. Specific and practical recommendations are made to address these limitations. Finally, the importance of improving the methodological quality of these member surveys is discussed.

TYPICAL SURVEY METHODOLOGY

For several reasons insurance companies usually contract with a vendor (e.g. a survey organization) to conduct many or all aspects of a sample survey of plan members. First, it is assumed that plan members will feel free to express their opinions to a neutral vendor without fearing personal impact upon their health care, especially if the survey assesses consumer satisfaction with the plan. Second, the insurer will not be able to identify which plan members are in the sample, thus ensuring confidentiality of sampled members. Third, the insurer may not have the required expertise and/or specialized resources to conduct all aspects of the sample survey.

One prototype survey procedure used by some insurers and vendors, as is or modified, is the CAHPS (Consumer Assessment of Health Plan Surveys) Survey and Reporting Kit (CAHPS 3.0, 2003). A consortium of public and private research organizations developed the CAHPS prototype with the support of AHRQ (Agency for Healthcare Research and Quality). Although developed at first specifically for assessment of consumer satisfaction with health care plans, the sampling methodology is used now for member surveys with purposes beyond consumer satisfaction. The CAHPS kit is comprehensive and includes data collection instruments in English and Spanish along with detailed suggestions for sampling plan strategies, response rate calculations, data analytic strategies (along with SAS programs) and reporting templates.

The sampling frame for these member surveys generally is developed from a list of all *members* of the health plan, although a list of *subscribers* may be used instead. One subscriber (e.g. an employee) may also cover his/her spouse and two children, yielding one subscriber but four members (two adults and two children). Other terminology is "policy-holder" for subscriber and "covered lives" for members. Some subscribers or members may be deleted from the list before finalizing the sampling frame, e.g. those who have not been in the plan continuously for at least six or twelve months. Children are excluded for surveys of adults only, and vice versa.

Once a sampling frame of members is finalized, the intent generally is to select a simple random sample of members from the frame. If the sampling frame contains a list of subscribers, generally a simple random sample of subscribers is selected, and CAHPS 3.0 recommends that the second stage of sampling select one member from the members associated with the selected subscriber. Because of respondent burden CAHPS 3.0 recommends that no more than one member per household be selected for any sample survey.

The insurer provides home address and telephone number for all plan members to the vendor. Usually a pre-notification letter is mailed to the sample member at his/her home address. One of the following four protocols is used for data collection, with the last two most common:

- Mailed, self-administered questionnaire;
- Telephone interview, often with CATI (computer assisted telephone interviewing);
- Mailed, self-administered questionnaire, with telephone interview for non-respondents;
- Telephone interview, with mailed, self-administered questionnaire for non-respondents.

The vendor attempts to correct an incorrect home address or telephone number in trying to locate a sample member. Upon contact with a sample member, eligibility for the survey is assessed. Eligibility usually is defined as currently covered by the health plan and may also include other specific characteristics, e.g. within a given age range.

CAHPS 3.0 recommends a minimum sample size of 300 completed telephone interviews or questionnaires per health plan. Recommended minimum response rates by CAHPS 3.0 are 60% for commercial populations and 50% for the Medicaid population. The survey response rate is defined as (number of completed returned questionnaires)/(number of members selected – number deceased – number ineligible). Response rates for these surveys often are not reported but range between 20% and 80% when reported (Fowler et al., 1999; Hays et al., 1999).

Data analysis typically is done with SAS, SPSS or a similar statistical package, using the sample members with completed questionnaires (or telephone interviews) as though they constitute a simple random sample from the population of inference. Analyses done by vendors or insurers often are limited to one-way or two-way percentage distributions. CAHPS 3.0 has guidelines and SAS programs for performing case-mix adjustment when comparing plans to each other, as well as SAS programs for computing indices for consumer satisfaction that are based on responses to several items.

Problems with Current Survey Methodology

A major problem with these surveys is frequent deviation from equal probability sampling of members coupled with unweighted analyses; this combination of sampling and analysis yields biased estimators (as shown later). Insurers and vendors may not realize that the sampling plan, as commonly implemented, yields an unequal probability sample of members. Note that CAHPS 3.0 does recommend a simple random sample of members. However, CAHPS 3.0 also states that only one person per "subscriber unit" should be selected for the sample so as to minimize respondent burden on a household. CAHPS 3.0 gives no explicit instructions on how to select a simple random sample with this constraint.

The following procedure seems commonly used to select a sample of adult members. Beginning with the list of all plan members, all subscriber units are identified. All children within subscriber units are excluded, as well as members who have not been with the plan the required length of time. For each subscriber unit with more than one adult, one adult from that unit is randomly (with equal probability) chosen to remain on the sampling frame. A simple random sample is selected from the resulting sampling frame. Possibly not realized by the insurer or vendor is that the obtained sample is not an equal probability sample (and hence not a simple random sample) of adult members. Members in subscriber units with x adults ($x > 1$) have a selection probability that is $1/x$ of the selection probability of members in subscriber units with one adult. That is, members of subscriber units of size one have a larger probability of coming into the sample, compared to members who are in subscriber units of size two or more.

A similar procedure for selecting a random sample of child members gives a smaller selection probability to children in subscriber units with more than one child, compared to children who are in subscriber units with only one child.

Deviation from equal probability sampling of elements (i.e. members) is common in sample surveys and may be used for many reasons, including reduction of respondent burden within a household. However, a problem may occur at the data analysis stage in these surveys when, typically, all analyses are unweighted. When unequal probability sampling is used, weighted analyses are required to obtain unbiased estimators, or approximately unbiased in the case of ratio estimators (Brogan, 1998). The analysis weight for a sample member is a function of that member's selection probability.

Comments have been made that better monitoring of the survey vendors is required because the sample of completed interviews or questionnaires often does not match the population of members (Brogan, 1999). Potential reasons for this "mismatch" are the deviation from equal probability sampling coupled with unweighted analyses (discussed here) as well as unit non-response (discussed below).

A second major problem is that location of and contact with sampled members frequently suffer from substantial non-response. Telephone is the preferred method to locate and personally contact sample members, but frequently only about 50% of the sample can be located and contacted by telephone. Often the insurer provides an incorrect home telephone number to the vendor. Even when the home telephone number is correct, the sample member may be away from home a lot or screen incoming calls.

The insurer's administrative database can have the wrong home telephone number for several reasons. Often the employee subscriber has failed to inform his/her employer of a change in home telephone number, or the employer may fail to inform the insurer of a change. Members of managed care plans, compared to indemnity plans, generally have less contact with their plans for financial reimbursement reasons and, thus, may have less opportunity or motivation to provide changes in home telephone number and address.

If the sample member cannot be located at the telephone number provided to the vendor, telephone directories, criss-cross directories (using home address) and other resources are used. However, these strategies do not work for unlisted telephone numbers. Sometimes a letter is sent to the home address provided by the plan, requesting the sample person to contact the vendor for a telephone interview. However, the home address of the sample person may not be up to date either.

When a mailed questionnaire is the first contact after the pre-notification letter, many sample persons may not return a completed questionnaire. The vendor becomes aware of some incorrect mailing addresses with questionnaires or pre-notification letters returned by the post office, and attempts are made to find a correct address. Telephone follow-up with mail questionnaire non-respondents suffers from the home telephone problems mentioned above.

A third common problem is a non-trivial non-eligibility rate. Some sample persons, once contacted, may no longer belong to the plan or may be deceased.

These sample persons are defined as ineligible for the survey; they simply drop out of the sample and do not negatively influence the survey response rate. However, financial resources are spent in determining the survey ineligibility of these sample members.

A fourth common problem is miscalculation (or non-calculation) of the survey response rate. CAHPS 3.0 recommends calculating the response rate as the percentage of completed interviews or questionnaires among sample persons who are eligible. However, vendors seem to give different interpretations to "sample persons who are eligible," and this can result in wide variation of reported response rates as well as suspiciously high response rates (Goyder et al., 2002). An unfortunately common response rate calculation seems to be the percentage of completed interviews among sample members who are located, contacted, eligible and consent to participate. Clearly, this is an overestimate of the survey response rate, especially with the major difficulty of locating and contacting sample persons.

A fifth common problem is that these surveys typically perform no adjustments for non-response; this is another reason for potentially biased estimators of population parameters. A "middle class bias" has been noted in many sample surveys, where those of higher SES are more likely to respond (Cohen & Duffy, 2002; Goyder et al., 2002). Further, response to health surveys, particularly among older persons, has been shown to be associated with better health (Cohen & Duffy, 2002).

Non-response adjustments are not guaranteed to or even necessarily expected to alleviate all potential biases from a low response rate. However, such adjustments can be useful in making the responding sample members match the selected sample on demographic or other characteristics. Similarly, these adjustments can make the weighted respondents match the population of members on demographic and other characteristics.

Since simple random sampling commonly is used or attempted in these surveys, a sixth potential problem is that subpopulations that comprise a small percentage of the population of inference typically are not represented in the sample with an adequate sample size for subpopulation analyses. For example, sample members who have frequent encounters with the health plan constitute a small percentage of the population and the sample. However, consumer satisfaction in this subpopulation may differ from those who have more limited interaction with the health plan during the past year. Insurers likely are interested in assessing consumer satisfaction not only for all members (or subscribers), but also for specific subpopulations. CAHPS 3.0 now addresses this issue in the context of providing an adequate sample size of child members who have a chronic health problem.

Suggested Improvements to Sampling Plan

The first suggestion is to make administrative data on members more timely and accurate. For example, when the intent is to sample current members, data collection costs could be reduced if members who have left the plan recently could be identified and eliminated from the sampling frame. The lag time for updating the administrative database to delete those who leave the plan may need to be shortened. Similarly, provision to the vendor of accurate home telephone numbers and addresses for all plan members could substantially reduce vendor time and costs in locating and contacting sampled members, as well as dramatically improve the survey response rate.

The second suggestion is to consider various sampling frames and sampling plans, taking into account the resulting weighting implications. It is not necessary to use only simple random sampling or equal probability sampling to obtain a probability sample of members. In fact, simple random sampling rarely is used in most sample surveys. Alternate sampling plans should be considered to meet specific survey objectives while recognizing practical constraints.

Consider the constraint that no more than one member from any subscriber unit be selected for the sample. The sampling frame could be constructed as a list of subscriber units rather than a list of members. A simple random sample of subscriber units could be selected. Then one sample member could be selected, with equal probability, from the eligible (e.g. by age) members of the subscriber unit. If there is only one eligible member within a selected subscriber unit, that member is selected with probability 1.0. This two-stage cluster sample results in an unequal probability sample of members, with members in subscriber units of size one having a higher selection probability.

Alternatively, the sample of subscriber units on the sampling frame could be selected with probability proportional to number of survey eligible (e.g. by age) members in the subscriber unit, i.e. pps (probability proportional to size) sampling. Then, one eligible member could be selected, with equal probability, from each selected subscriber unit. In this two-stage cluster sampling plan an equal probability sample of members is obtained (assuming the size measure is accurate for each subscriber unit).

CAHPS 3.0 briefly mentions two-stage sampling strategies as suggested above but gives no specific guidelines other than to consult with a sample survey statistician. Also, Gallagher et al. (1999) discuss the intriguing problem of having only subscriber information available and then using mail administration to perform subsampling among members nested within subscriber units.

Even though it may be desired to avoid the selection of two or more members per subscriber unit, a sampling plan that gives a positive, but small, probability to this event may be quite acceptable as a trade-off for a simpler sampling plan.

The third suggestion is to oversample subpopulations of interest. It is likely that consumer satisfaction or other population parameters of interest vary among subpopulations, e.g. by demographic characteristics or health status. If one subpopulation of interest constitutes a small percentage of all members, then with equal probability sampling of members this subpopulation will constitute a small percentage within the sample. In order to increase the subpopulation's sample size, an oversampling strategy can be used; i.e. members of the subpopulation are given a higher probability of being selected for the sample.

Frequent users of health services, as a subpopulation example, could be defined by number of medical visits during the past 12 months and/or by particular medical diagnoses. Information may be available in the insurer's administrative database or computerized medical records to identify members who belong to the subpopulation.

The easiest strategy for oversampling is to stratify all members into the subpopulations of interest (e.g. frequent and infrequent users) and then use a higher sampling fraction in the smaller subpopulation(s), i.e. stratified random sampling with disproportionate allocation (of the total sample size). However, this technique might result in more than one member being selected within a subscriber unit. A second sampling technique is two stage cluster sampling, where the primary sampling unit (PSU) is subscriber unit, followed by stratification of members of sampled PSUs into subpopulations before second stage sampling. Control techniques can be used to minimize the probability of selecting more than one member per subscriber unit (Whitmore et al., 1998). Use of any of these sampling strategies yields an unequal probability sample of members.

CAHPS 3.0 suggests a stratified random sampling plan for child plan members where the population of child members is stratified by likely presence/absence of chronic condition(s). The stratum with likely presence is oversampled. The stratification variable here does not have perfect sensitivity (for chronicity) or specificity (for non-chronicity). Thus, when the survey data determine that a sampled child has chronicity, that child could have been sampled from either the likely presence or likely absence stratum. In this situation, the stratification variable (likely presence/absence) is used to keep track of the stratum from which the sample child was selected (for variance estimation purposes), but the analytical variable of interest is the actual presence/absence of chronicity as determined from the respondents' survey data.

Note that there may be confidentiality and/or ethical issues to consider for stratified random sampling since the value of the stratification variable needs to be known (or guessed at) for each population member on the sampling frame. Typically the information for stratification of plan members would come from administrative, utilization or medical databases about the members.

Suggested Improvements to Field Operations

First, improved procedures clearly need to be developed for locating and contacting sample persons, although addressing sampling frame problems mentioned in the previous section will decrease problems at these stages. Once contact is made and eligibility is determined, it seems that a majority of sample members agree to participate and complete a telephone interview or self-administered questionnaire.

It is becoming more difficult in the U.S. to conduct telephone surveys; their response rates have declined over the past several years (Brogan et al., 2001; Groves & Couper, 1998). Most telephone surveys use some variation of random digit dialing to select a probability sample of households, followed by a probability selection of one or more persons within the household and a telephone interview with each selected person. In general, the content of the telephone interview is not salient to the randomly selected person. Further, many householders have negative reactions to legitimate telephone survey researchers because of previous experiences with telemarketing. However, these general factors should have a small impact on the telephone protocol for surveys of health plan members. First, a satisfaction or health-related survey is likely to be a salient topic to the sample member, and, two, the sample member is asked for specifically by name, not as a result of some random selection once the household is reached.

Improved methods for locating and contacting sample members most likely depend upon whether the surveyed population is commercial, Medicaid or Medicare. The commercial population often has insurance coverage through an employer, and the plan knows who the employer purchaser is for any given sample member. Perhaps the sample person or the subscriber associated with the sample person could be located at work in order to obtain a correct telephone number and/or address for a later contact at home. The Medicaid population is insured through government, so other strategies may be needed here; this has been a difficult problem (Brown et al., 1999; Gibson et al., 1999). In general, the Medicare population seems easier to locate and contact since most of them are not in the labor force and perhaps more likely to be at home; they also tend to move less frequently than younger people.

A second suggestion is to calculate stage-specific response rates and then an overall survey response rate. Stage-specific response rates alert the insurer and vendor to the stages most susceptible to non-response, leading to efforts to improve procedures at these stages. Further, the overall survey response rate can be calculated easily as the product of the stage-specific response rates. Four stages are recommended for a typical health plan member survey, based on generic suggestions from Lessler and Kalsbeek (1992) and from AAPOR (2003).

(1) LOCATE. Determine that the sample member can be reached at a home telephone number and/or address; this information can be obtained by proxy. From the total selected sample of n, L are located and $(n - L)$ are not located. The "Locate Response Rate" (L/n) is the proportion of all sample members who are located.
(2) CONTACT/SCREEN. Make contact with the located sample member (or acceptable proxy informant) and screen for survey eligibility, typically defined as alive, current and/or 6–12 month member of the plan and any other requirements such as age. Of the L located, D are found to be deceased (via proxy), E are found to be eligible, NE are found to be alive but not eligible and the remainder $(L - D - E - \mathrm{NE})$ are not contacted and assessed for eligibility. The "Contact/Screen Response Rate," defined as $((E + \mathrm{NE} + D)/L)$, is the proportion of all located sample persons who are contacted (in person or via acceptable proxy) and screened for eligibility. Another rate of interest may be the "Eligibility Rate" $(E/(E + \mathrm{NE} + D))$, i.e. the proportion of contacted/screened persons who are eligible. This rate, hopefully near 1.0, assesses the accuracy and resulting efficiency of the sampling frame.
(3) PARTICIPATE. Obtain consent from eligible sample member to be interviewed or complete a self-administered questionnaire. Of the E eligibles, P agree to participate and $(E - P)$ do not agree. The "Participation Rate" (P/E) is the proportion of eligible sample members who agree to participate in the survey.
(4) COMPLETE. Obtain a "completed" telephone interview or self-administered questionnaire from the participating member. Of the P who agree to participate, C complete an interview or questionnaire and $(P - C)$ do not. The "Completion Rate" (C/P) is the proportion of participating sample members who complete an interview or self-administered questionnaire.

The survey response rate (SRR) is the product of the four response rates above, i.e. the rates for Locate, Contact/Screen, Participation and Completion:

$$\mathrm{SRR} = \left(\frac{L}{n}\right)\left(\frac{E + \mathrm{NE} + D}{L}\right)\left(\frac{P}{E}\right)\left(\frac{C}{P}\right).$$

Clearly, the "Completion Rate" (C/P) is an overestimate of the survey response rate SRR. Further, SRR is bounded above by the "Locate Rate" (L/n), often the lowest response rate among the four defined above.

Suggested Improvements to Data Analyses

The first suggestion is to use weighting procedures to reflect the sampling plan that was actually used (most likely unequal probability sampling of members). A weighted analysis is required for unbiased (or nearly unbiased) estimators of population parameters whenever unequal probability sampling is used. The initial sampling weight for each sample member is the inverse of the selection probability.

To illustrate the magnitude of bias in using unweighted analyses with an unequal probability sample, consider a population of N adult members of a health plan who are eligible for the survey. Let N_1 be the number of members in subscriber units of size one; let N_2 be the number of members in subscriber units of size two, i.e. there are $N_2/2$ subscriber units of size two. For simplicity assume there are no subscriber units of size larger than two, i.e. $N = N_1 + N_2$. Let M be the population mean of some variable for the N members, e.g. mean satisfaction rating or proportion who are very satisfied. It is desired to estimate the population parameter M. Let M_1 and M_2 be the subpopulation mean of this same satisfaction variable for the N_1 members and N_2 members, respectively. Let $p = N_1/N$ be the proportion of members who are in subscriber units of size one. Then the overall mean M can be represented as:

$$M = pM_1 + (1 - p)M_2.$$

Now assume the commonly used sampling plan discussed earlier. That is, the sampling frame is prepared from the list of N members by removing, at random, one member from each of the subscriber units of size two. The sampling frame then contains $(N_1 + N_2/2)$ members. The mean of the satisfaction variable for the N_1 members on the sampling frame is still M_1; the mean of the satisfaction variable for the $N_2/2$ members on the sampling frame is a random variable with expectation M_2. The proportion of members on the sampling frame who are in subscriber units of size one is (via algebra) $2p/(1 + p)$; the proportion who are in subscriber units of size two is $(1 - p)/(1 + p)$. Thus, the expected mean of the satisfaction variable for all members on the sampling frame is M^*, where

$$M^* = \frac{2M_1p}{1+p} + \frac{M_2(1-p)}{1+p}.$$

Note that $M^* \neq M$, where M is the mean of the satisfaction variable over all N members and M^* is the expected value of the mean of the satisfaction variable over all possible sampling frames that could be constructed by deleting one member from each subscriber unit of size two. Thus, in order to estimate the population parameter M, a probability sample is taken from a sampling frame with expected mean satisfaction score of M^*; this clearly is not desirable.

Assume now a simple random sample from the sampling frame, with a 100% response rate. Then the two-stage expectation (first over selecting a sample from the sampling frame and then over construction of the sampling frame from the population of all members) of the unweighted sample mean is the expected mean of the sampling frame, i.e. M^*. Thus, the bias of the unweighted sample mean is given by $(M^* - M)$. After some algebra, this bias can be represented as:

$$\text{Bias\_1} = M^* - M = \frac{p(1-p)(M_1 - M_2)}{1+p}.$$

Bias_1 is zero under any one of three conditions below:

(1) $p = 0$, i.e. all N members are in subscriber units of size two. Then the population of $N = N_2$ members has mean $M = M_2$, and the sampling frame of $N_2/2$ members has expected mean M^*, where $M^* = M_2 = M$.
(2) $p = 1$, i.e. all N members are in subscriber units of size one. The population and sampling frame both contain $N = N_1$ members, and both have the same mean $M = M_1$.
(3) $M_1 = M_2$, i.e. members in subscriber units of varying size (1 or 2) have the same mean value of the satisfaction variable.

Since it is unlikely in practice that any of these three conditions hold, this common sampling and analytic procedure yields a biased estimator of the population mean M. The magnitude of Bias_1 increases as M_1 and M_2 deviate from each other. With respect to p, Bias_1 is maximized when $p = 0.414$. In this situation, 41% of the N members in the population are in subscriber units of size one, but when the sampling frame is constructed, these types of members comprise 71% of the sampling frame. Clearly the population and the sampling frame are quite different on the distribution of members by size of subscriber units.

Bias formulas could be derived with different configurations of the population by size of subscriber units, e.g. a mixture of subscriber units of size one, two and three. The findings will be similar; i.e. this common sampling procedure to minimize respondent burden within a subscriber unit, combined with unweighted analyses, yields biased estimators of population parameters.

The bias for the unweighted mean estimator (derived above) can be eliminated by using a weighted mean estimator. The value of the sampling weight for a sampled member would be

$$\frac{(N_1 + N_2/2)}{n} \quad \text{and} \quad \frac{2(N_1 + N_2/2)}{n}$$

for members in subscriber units of size one and two, respectively. Note that members in subscriber units of size two have a larger value of the sampling weight since they have a smaller probability of being selected, compared to members in subscriber units of size one. Note also that use of these sampling weights based only on selection probability assumes a survey response rate of 100% (since no adjustments for non-response are made).

Standard statistical packages such as SAS or SPSS can be used to yield weighted point estimates (e.g. a weighted mean); a WEIGHT statement is used where the variable on the weight statement is the sampling weight variable. If standard errors are desired for point estimates or other analyses, procedures in standard statistical packages that assume simple random sampling will not yield appropriate estimates for sampling variability (Brogan, 1998). A specialized statistical software package for sample survey data can be used, e.g. SUDAAN or WESVAR. Alternatively, some standard statistical packages include specialized procedures to analyze sample survey data (e.g. STATA, SAS, EPI-INFO).

A second suggestion is to consider weighting procedures to adjust for unit non-response. Even if the overall survey response rate is increased beyond current levels by adopting some of the suggestions above, unit (member) non-response will still occur at each of the four stages. Ignoring the non-response in data analysis may also contribute further bias to estimators of population parameters. The insurer's administrative database contains demographic and other characteristics about population and sample members (e.g. age, gender, geographic residence, type of plan, subscriber unit size). Thus, one can ascertain whether the sample members who respond (respondents) differ on these characteristics from the sample members who do not respond (non-respondents), overall and/or at the four specific response stages discussed in the previous section. Alternatively, one can compare the weighted (by inverse of selection probability) respondents to the entire population of members on these characteristics to determine potential response bias. If respondents do differ from non-respondents or if the weighted respondents differ from the population of members on characteristics likely related to the analysis variables, then standard complex sample survey procedures can be used to differentially weight the respondents to "adjust" for unit non-response (Lessler & Kalsbeek, 1992; Lohr, 1999).

To illustrate how non-response can contribute to bias in estimators, let the population, sampling frame, and sampling procedure be defined as in the previous section. However, now let r_1 and r_2 be the proportion of sample members who respond among those in subscriber units of size one and two, respectively. For simplicity, assume that the respondents and non-respondents, within subscriber units of a given size (one or two), have the same mean on the variable of interest. Under these conditions and using expectations as earlier, it can be shown that the bias of the unweighted sample mean, based on the sample respondents, is given by Bias_2, where

$$\text{Bias\_2} = \frac{p(1-p)(2r_1 - r_2)(M_1 - M_2)}{2r_1 p + r_2(1-p)}.$$

Note that Bias_2 reduces to the previous Bias_1 if $r_1 = r_2 = 1.0$, i.e. a 100% response rate. Any of the three conditions discussed earlier that make Bias_1 $= 0$ also will make Bias_2 $= 0$. Bias_2 also equals zero if $r_2 = 2r_1$; this condition, in effect, "cancels out" the over-representation of members in single subscriber units on the sampling frame by giving a smaller response rate to these members in the sample. Any of the conditions that make Bias_2 $= 0$ are unlikely to happen in practice. Thus, non-response also contributes to the bias of the unweighted estimator used in the common sampling plan that avoids selection of more than one member per subscriber unit.

A more realistic assumption is that, within subscriber units of a given size, the mean for respondents is not the same as the mean for non-respondents. The bias of the unweighted estimator then would be a function of the terms in Bias_2 above, but also a function of the difference between the means for respondents and non-respondents within subscriber units of a given size.

The third suggestion is to conduct analytical as well as descriptive analyses. Vendor or insurer conducted analyses of sample surveys of health plan members, particularly consumer satisfaction surveys, tend to be primarily descriptive. These analyses focus on the estimation of percentage of members who have a particular characteristic, e.g. some specific health characteristic or consumer satisfaction with various dimensions of their health care. Sometimes analyses are conducted for subpopulations defined by demographic characteristics. However, these survey data provide opportunities to do more analytical or model building analyses (Zaslavsky et al., 2002; Zaslavsky & Cleary, 2002; Zhan et al., 2002). For example, analyses could identify determinants of consumer satisfaction, as well as which components of satisfaction are most strongly related to overall satisfaction; this could indicate targeted areas for plan improvement.

DISCUSSION

All of the suggestions above for survey improvement are standard procedures in complex sample survey methodology (Lohr, 1999). However, health care insurers and their survey vendors do not appear to be systematically using these procedures in health plan member surveys. Although quantitatively trained professionals (e.g. epidemiologists, biostatisticians, psychologists, sociologists, economists) typically are involved in these surveys, there are practical problems at the interface between complex sample survey methodology and these health plan member surveys. Many quantitatively trained health research professionals are not familiar with the specialized and technical aspects of designing and analyzing complex sample surveys.

Sample survey statisticians, as consultants, could advise insurers and vendors on implementation of the suggestions made above, e.g. sampling plans that deviate from simple random sampling, appropriate weighting of survey data, unit non-response adjustments, and analytical strategies for weighted data. Some of these topics are mentioned in CAHPS 3.0 but not in enough detail for a typical vendor or insurer to implement without technical consultation. In fact, the CAHPS 3.0 documentation frequently recommends consultation with a sample survey statistician.

The bias formulas presented above have not been published before (to the author's knowledge). They illustrate the magnitude of potential bias for a commonly used sampling plan that tries to minimize respondent burden within subscriber unit, coupled with an unweighted analysis. This bias can be eliminated, or at least reduced, by appropriate weighted analyses where the value of the sampling weight for each respondent is a function of the selection probability and adjustments for unit non-response.

These suggestions for improving surveys of health plan members may incur additional cost, but this may be offset, at least partially, by reduced fielding costs, e.g. in locating sample members. But more important, multiple implementation aspects of these surveys raise concern whether the current data analyses yield valid inferences about the health plans or their members. Survey data on consumer satisfaction and health status of members contribute to major decisions about the health plan: e.g. (1) whether the plan is accredited; (2) whether quality medical care is being delivered to consumers; (3) whether the plan needs to change its systems for delivering health care; and (4) how to market the plan to employers, governments and consumers. To make these crucial decisions based on questionable data sources likely is much more expensive in terms of potentially wrong decisions than the cost of improving these important surveys.

ACKNOWLEDGMENTS

This manuscript is an extension of a presentation on the same topic on September 22nd, 1999 in Edinburgh, Scotland at the 3rd International Conference of

Association for Survey Computing. Some ideas in this manuscript are based on a luncheon round table discussion on August 9th, 1999 on "Surveys of Consumer Satisfaction with Health Plans," moderated by Donna Brogan, at the Joint Statistical Meetings (JSM) in Baltimore, MD. I thank Dr. Linda Fischer of Washington, D.C. for comments on an earlier version of this manuscript.

REFERENCES

AAPOR (2003). Standard Definitions: Final Dispositions of Case Codes and Outcome Rates for Surveys (RDD Telephone Surveys, In-Person Household Surveys, Mail Surveys of Specifically Named Persons). http://www.aapor.org/default.asp?page=survey_methods/standards_and_best_practices/standard_definitions#aapor

Brogan, D. (1998). Software for sample survey data: Misuse of standard packages. In: P. Armitage & T. Colton (Eds), *Encyclopedia of Biostatistics* (Vol. 5, pp. 4167–4174). New York, NY: Wiley.

Brogan, D. (1999). Surveys of consumer satisfaction with health plans. Luncheon round table discussion on August 9th, 1999 at joint statistical meetings (JSM) in Baltimore, MD.

Brogan, D. J., Denniston, M. M., Liff, J. M., Flagg, E. W., Coates, R. J., & Brinton, L. A. (2001). Comparison of telephone and area sampling: Response rates and within household non-coverage. *American Journal of Epidemiology*, *153*(11), 1119–1127.

Brown, J. A., Nederend, S. E., Hays, R. D., Short, P. F., & Farley, D. O. (1999). Special issues in assessing care of Medicaid recipients. *Medical Care*, *37*(Suppl. 3), MS79–MS88.

CAHPS (2003). CAHPS survey and reporting kit 2002 with CAHPS 3.0 questionnaires. www.cahps-sun.org and www.ahcpr.gov/qual/cahpsix.htm

Cohen, G., & Duffy, J. C. (2002). Are non-respondents to health surveys less healthy than respondents? *Journal of Official Statistics*, *18*(1), 13–23.

Fowler, F. J., Gallagher, P. M., & Nederend, S. E. (1999). Comparing telephone and mail responses to the CAHPS survey instrument. *Medical Care*, *37*(Suppl. 3), MS41–MS49.

Gallagher, P. M., Fowler, F. J., & Stringfellow, V. L. (1999). Respondent selection by mail: Obtaining probability samples of health plan enrollees. *Medical Care*, *37*(Suppl. 3), MS50–MS58.

Gibson, P. J., Koepsell, T. D., Diehr, P., & Hale, C. (1999). Increasing response rates for mailed surveys of Medicaid clients and other low-income populations. *American Journal of Epidemiology*, *194*(11), 1057–1062.

Goyder, J., Warriner, K., & Miller, S. (2002). Evaluating socio-economic status (SES) bias in survey non-response. *Journal of Official Statistics*, *18*(1), 1–11.

Groves, R. M., & Couper, M. P. (1998). *Non-response in household interview surveys*. New York, NY: Wiley.

Hays, R. D., Shaul, J. A., Williams, V. S. L., Lubalin, J. S., Harris-Kojetin, L. D., Sweeny, S. F., & Cleary, P. D. (1999). Psychometric properties of the CAHPS 1.0 survey measures. *Medical Care*, *37*(Suppl. 3), MS22–MS31.

Lessler, J. T., & Kalsbeek, W. D. (1992). *Non-sampling error in surveys*. New York, NY: Wiley (Chaps 6–8).

Lohr, S. L. (1999). *Sampling: Design and analysis*. Pacific Grove, CA: Duxbury Press, Brooks/Cole Publishing.

Whitmore, R. B., Folsom, R. E., Burkheimer, G. J., & Wheeless, S. C. (1998). Within-household sampling of multiple target groups in computer-assisted telephone surveys. *Journal of Official Statistics*, *4*(4), 299–305.

Zhan, C., Sangi, J., Meyer, G. S., & Zaslavsky, A. M. (2002). Consumer assessments of care for children and adults in health plans: How do they compare? *Medical Care*, *40*(2), 145–154.

Zaslavsky, A. M., & Cleary, P. D. (2002). Dimensions of plan performance for sick and healthy members on the consumer assessments of health plans study 2.0 survey. *Medical Care*, *40*(10), 951–964.

Zaslavsky, A. M., Zaborski, L. B., & Cleary, P. D. (2002, June). Factors affecting response rates to the Consumer Assessment of Health Plans Study survey. *Medical Care*, *40*(6), 485–499.

ORGANIZATIONAL VARIATION IN THE MANAGED CARE INDUSTRY IN THE 1990S: IMPLICATIONS FOR INSTITUTIONAL CHANGE

Denise Anthony and Jane Banaszak-Holl

ABSTRACT

Despite continuing debate about costs and benefits, managed care became an integral part of the health care sector during the 1990s. In this paper, we examine the organizational and practice variation in the managed care industry at two points in the 1990s using a national census of organizations operating in those years. We use a definition of managed care that captures the increased diversity within the industry while still distinguishing it from traditional indemnity, fee-for-service care. We draw on institutional theory to begin to formulate a framework for understanding why certain organizational forms and practices emerged when and where they did.

INTRODUCTION

In the last quarter of the 20th century, the social organization of health care delivery in the U.S. changed dramatically with the introduction and growth of managed care (MC). In a nutshell, MC is comprised of both organizations and

Reorganizing Health Care Delivery Systems: Problems of Managed Care and Other Models of Health Care Delivery
Research in the Sociology of Health Care, Volume 21, 21–38

ISSN: 0275-4959/doi:10.1016/S0275-4959(03)21002-4

organizational practices that manage as well as finance the delivery of health care. MC organizations are prevalent in all health care markets (Wholey et al., 1992) and managed care is the dominant form of health coverage for privately-insured individuals (Gabel, 1999; Gabel et al., 1994). MC has achieved what neo-institutional theorists label a "taken-for-granted" status (DiMaggio & Powell, 1983; Meyer & Rowan, 1977; Zucker, 1977). In other words, MC is now a social institution in the United States.

During its early development in the 1970s and 1980s, MC was synonymous with the organizational form of group-staff model health maintenance organizations or HMOs (e.g. Group Health of Puget Sound and Kaiser-Permanente), which consisted of a clear and finite set of practices through which health care was managed: physicians paid by salary or "capitation" (i.e. annual lump-sum pre-payment) with primary care gatekeepers and frequently, utilization review (Miller & Luft, 1994; Strang, 1995). These and other features of MC were a direct challenge to the dominant paradigm of fee-for-service (FFS) medicine, but had little impact on the structure of health care initially. It was not until the cost-pressures from dramatic health care inflation in the 1980s, coupled with an economic recession, that the "logic" of FFS medicine was threatened by the alternative logic offered by MC. It was at that time, in the late 1980s, that states began to create regulatory incentives to promote MC (Strang, 1995; Strang & Bradburn, 2001; see also Light, 2001).

By the early 1990s, the paradigm shift away from FFS health care toward MC was in full swing, and by the mid-1990s MC had replaced FFS as the dominant organizing principle in health care insurance (Gabel et al., 1994), with widespread implementation of MC practices such as capitation, gatekeeping and utilization review. The widespread diffusion of MC forms and practices might appear to be the end of a story about the institutionalization of MC in health care. Indeed, the vast majority of analyses of MC over the past twenty years have been concerned with how MC *differs* from FFS health care, not how MC itself varies. We, however, take this period as our point of departure. As we show in this paper, the extent of organizational variation within MC indicates that the process of institutional change in health care financing and delivery is still very much underway. Scholars have described this as a period of profound institutional change in the health care industry (Scott et al., 2000; see also Mechanic, 2002). Though all forms of MC differ in important ways from FFS health care, the variation within MC has implications for understanding current and future effects of health care delivery for patients, for providers and for society as a whole.

In this paper, we explore the organizational composition of the managed care industry by describing the demographic distribution of organizational forms and practices across the U.S. at two points during the 1990s. Our goal here is threefold: (1) to determine the extent of organizational variation in managed care during the period of institutionalization in the 1990s; (2) to compose a new working definition of managed care that captures the increased diversity within the industry while still distinguishing it from traditional fee-for-service health care; and (3) to begin to formulate a framework for understanding why certain forms and practices emerged they did, by drawing on theories of institutional and organizational change. We raise more questions than we answer in this paper, in part because our primary goal is to illustrate the variation within MC and discuss its implications, rather than to fully analyze why we see the variation we do. MC, however, offers an opportunity to both illustrate and further develop theoretical models of institutional change, and this study is the first part of a larger project that will further analyze the dynamics of change within this industry.

INSTITUTIONAL CHANGE

Despite the increasing prevalence of MC throughout the 1990s, there is a great deal of variation in management rules and practices, while the prevalence and concentration of different forms of MC organizations varies considerably across local markets. As we will show below, by the end of the 1990s, the initially quintessential form of MC, the staff-model HMO had declined significantly, while new organizational forms, such as the individual practice association (IPA) and the preferred provider organization (PPO), emerged and grew rapidly throughout the decade. Thus, ironically, during the period of apparent institutionalization in which MC became the predominant form of healthcare delivery, the organizational features of managed care became more rather than less diverse.

Institutionalization often implies uniformity because it is a process through which institutions, resilient and stable social structures, are formed. As Scott (2001) makes clear, however, institutions operate at multiple levels as well as via multiple mechanisms (see also Campbell, 1997). Institutional change is a broad set of processes entailing both the adoption and evolution of new organizational forms and practices (what is often termed "institutionalization," DiMaggio & Powell, 1983), as well as the discontinuation or abandonment of other forms and practices (termed "deinstitutionalization," Oliver, 1992) (Jepperson, 1991). So while "changes in practice co-evolve with changes in legitimating logics" during institutional change (Scott, 2001, p. 190), they do not do so uniformly. Organizational change may begin with uncertainty about what practices an organization should adopt as previous norms and practices are deinstitutionalized but new standards are not yet established (Oliver, 1992; Strang & Soule, 1998). Organizations may innovate new rules or practices over a relatively short period of time, only to discard or radically alter those procedures if they do not become the institutionalized practices in the field (Dowell & Swaminathan, 2000). Innovation itself can be a catalyst for further organizational innovation, leading to increasing, rather than decreasing rates of organizational change (Greve & Taylor, 2000). Alternatively, organizations may adopt new practices only after they have become commonplace among peer organizations (DiMaggio & Powell, 1983), or mandated from the state or some other regulatory body (Edelman, 1992; Fligstein, 1990).

Even under strong institutional pressure, organizational variation may increase (Edelman, 1992; Lounsbury, 2001; Ruef & Scott, 1998). Organizational variation in response to normative and/or regulative pressures occurs partly because ecological and competitive pressures affect organizational responses to the institutional environment (Dacin, 1997; Haveman & Rao, 1997) and more specifically, because competitive and institutional environments vary in intensity at the local level (Hannan et al., 1995; Wade et al., 1998). At the same time, differential access to resources across firms leads to organizational variation in response to institutional pressures (Lounsbury, 2001; Suchman, 1995). Organizational practices inconsistent with dominant institutional logics, however, will not be readily incorporated across organizations (Biggart & Guillen, 1999).

Recognizing that institutions are multi-layered, and that institutional change is a multi-dimensional process, makes the increasing organizational diversity in the MC industry less perplexing because uniformity at one level may hide diversity at other levels. Further, it suggests that examining institutional change in the MC industry requires attention to multiple aspects of the organizational entities comprising it (Ruef, 2000; Scott et al., 2000). Figure 1 very simply illustrates institutional change across multiple organizational dimensions, including *institutional logics*, the broad organizing principles that indicate (and define) what is considered appropriate, normal and reasonable for organizations (Biggart & Guillen, 1999; Friedland & Alford, 1991); *organizational forms*, the set of authority structures and technological systems used across a population of organizations (Hannon & Freeman, 1989; Stinchcombe, 1965); and *organizational practices*, the rules, guidelines and routines that govern organizational tasks (DiMaggio & Powell, 1991; Nelson & Winter, 1982).

Organizational change may occur first on any dimension. If change at one level leads to significant changes at the other two levels, it results in institutional change. In their historical review of long-term care in the United States, for example, Kitchener and Harrington (2003) show that institutional change in the long-term care sector requires multi-dimensional change in *care practices* new *organizational forms* and *societal norms* of how to treat the elderly. Alternatively, change along one organizational dimensions may lead to no further change at the other levels,

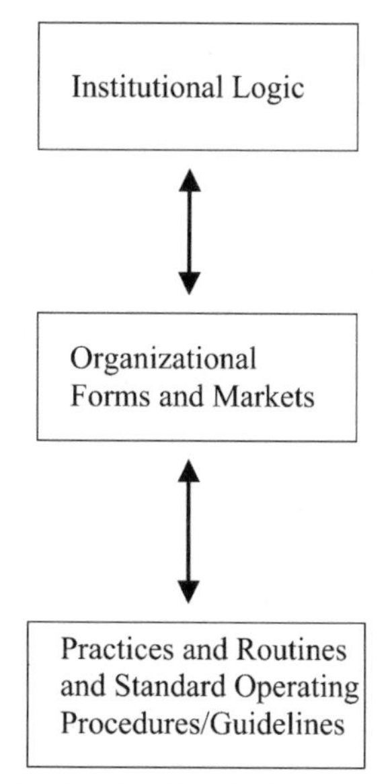

Fig. 1. Levels of Institutional Change.

resulting in organizational change but not significant institutional change. Many of the more interesting dynamics in organizational and institutional change occur because change across dimensions is not uni-directional and may occur at different speeds. A closer study of variation in MC can contribute to our understanding of the dynamics of institutional change across levels of analysis and can address many questions regarding the future of the health care industry.

Diversity in Managed Care Organizational Forms and Practices

From an institutional perspective, the "logic" of MC is the *idea* that both the clinical practice and the financing of health care should be managed together, typically by insurance plans. As noted above, this was a direct challenge to the logic of FFS medicine, in which financing and clinical practice are managed separately. Under FFS, insurance plans paid fees charged by physicians who provided clinical care; physicians did not share the risk of health insurance, and physicians managed clinical care via professional rules, relationships and norms. Under MC, physicians are expected to share risk as well as submit to oversight of their clinical practice by insurance plans. The stark difference in organizing principles alone has caused deep-seated resistance to the institutionalization of MC (Mechanic, 1996, 2002).

When institutions are indeterminate, such as when a dominant logic is challenged, or multiple logics exist, as is currently the case in health care, there is uncertainty regarding appropriate organizational forms and practices. Currently, HMOs and PPOs are the two major organizational forms of MC, but they differ significantly in the practices they use to pay providers and to manage care delivery (Gold et al., 1995). Previous research has found substantial differences in the survival and growth of different types of HMOs and of PPOs (Christianson et al., 1991; Gold et al., 1995; Wholey et al., 1992; Wholey & Burns, 1993). A substantial literature exists on the practices and historical trends among HMOs, including research on physician payment (Sleeper et al., 1998), insurance premium structures (Wholey et al., 1995), organizational forms (Wholey & Burns, 1993) and state oversight of HMOs (Christianson et al., 1991). In contrast, though a number of studies analyze the organizational features of PPOs (e.g. Dalton, 1987; Gold et al., 1995; Gold & Hurley, 1997), and the effects of PPOs on health costs and utilization (e.g. Hellinger, 1995; Smith, 1997), PPOs have been relatively unstudied despite their increasing prevalence in the last ten years. Here, we compare the ecological trends in both HMOs and PPOs simultaneously to determine how the larger population of MC firms more generally has evolved. Consequently, we compare the prevalence and practices of both PPO and HMO forms of managed care in 1993 and 1998, and in so doing, explore the relative differences and similarities within and between these two organizational forms.

Data

We examine MC organizational forms and practices using data reported in the Medical Economics Company (MEC) HMO/PPO Directory for the years 1993 and 1998. The directories provide comprehensive information on the organizational characteristics for all MC health plans operating in each state in the U.S. Comparison to the more familiar InterStudy HMO (1998) and PPO Directories (2001), identify similar numbers of organizations for comparable years. The MEC Directories provide detailed information on specific organizational practices, including oversight and provider payment mechanisms structures. While cross-sectional, these data provide valuable first insights into population level changes in MC during the last decade.

Health plans are categorized as HMO or PPO. HMOs are defined as health plans offering prepaid, comprehensive health coverage for both hospital and physician services, in which members are required to use participating providers and are enrolled for a specified period of time. PPOs are defined as plans in which beneficiaries receive care from a selected panel of providers who agree to some form of discounted fee schedule when contracting with the PPO.

Model types within plans are categorized as one of three possible types. The *group or staff* model is one in which a contracted or salaried physician group provides health services to a health plan's members. An Independent Practice Association (*IPA*) *or network* model is a plan that contracts directly with one or more independent physician practices, of which the practices may be all or some combination of solo, single-specialty or multi-specialty group practices. Models classified as *other* are plans that did not specify a model type.

We document differences in a number of organizational-level characteristics of MC plans including whether the organization is for-profit or is nationally accredited. For the 1998 data, we also have information regarding whether the organization issues a report card.

A central focus of this paper is to understand not only the variation in organizational forms and characteristics, but also the variation in practices across and within forms. The directories provide information on whether the plan implements a variety of MC practices, including: utilization review, required second surgical opinions and case management, whether patients are required to select a primary care provider, as well as provider-payment methods (i.e. salary, capitation or discounted-fee-for-service). Data on payment methods are available for 1998 only. The directories list whether plans respond "yes" to using any of these practices. The data we report are percentages of plans that reported "yes," compared to all other plans, including both those that said "no," as well as those that reported "not applicable."

FINDINGS

Describing the MC Organization Population

Table 1 shows that the number of MC organizations continued to increase during the 1990s, although the distribution of MC forms changed during the decade. Overall, HMOs were still the predominant organizational form in 1998, but PPOs increased dramatically in both number and as a proportion of all MC organizations between 1993 and 1998. Comparing the general categories of PPO to HMO (shaded rows in Table 1), the number of PPOs increased by 40% while HMOs increased 21% between 1993 and 1998. In 1998, PPOs comprised 46% of all MC organizations compared to about 39% in 1993. PPOs also have much higher numbers of enrollees than HMOs.[1]

Table 1. Characteristics of Different Forms of Managed Care Organizations in the 1990s.

	Mean Year Started	Total Number Organizations		% Change	Mean Number Enrollees	
		1993	1998	1993–1998	1993	1998
HMO	1981	517	654	+21	89,300	521,800
Group/Staff	1976	93	64	−45	107,178	498,405
IPA/Network	1982	412	528	+22	78,047	523,227
Other	1983	12	62	+81	336,235	553,453
PPO	1984	335	561	+40	249,800	794,600
IPA/Network	1983	238	438	+46	250,882	865,825
Other	1985	97	123	+21	247,157	357,767

Source: Medical Economics Company HMO/PPO Directory, 1993 and 1998.

Comparing in more detail by type of HMO or PPO shows a more complex picture (see the unshaded rows in Table 1). The earliest form of managed care, the HMO-Group/Staff model, declined dramatically during the 1990s, although newer types of HMOs increased in number. Note that the IPA/Network model of both HMOs and PPOs is the predominant type within each form, in both number of organizations and number of enrollees.

Table 2 shows that the general pattern of declines in HMO-Group/Staff models, and increases in IPA/Network models in both forms, as well as the dramatic increase in PPOs overall, holds true for every region across the country. Consistent with the idea that MC is becoming institutionalized, we see convergence in the

Table 2. Regional Distribution of Managed Care Organizations.

	North East		South		Mid-West		West	
	1993	1998	1993	1998	1993	1998	1993	1998
HMO	68.1%	52.2%	62.1%	56.3%	64.9%	56.8%	51.4%	49.9%
Group/Staff	12.8	4.5	13.3	3.9	11.2	6.9	7.8	5.5
IPA/Network	51.8	40.2	48.3	47.3	52.6	45.3	42.4	40.3
Other	3.5	7.6	0.5	5.1	1.2	4.5	1.2	4.0
PPO	31.9%	47.8%	37.9%	43.7%	35.1%	43.2%	48.6%	50.1%
IPA/Network	20.6	33.5	26.1	34.4	26.7	34.5	34.6	40.6
Other	11.3	14.3	11.8	9.3	8.4	8.7	14.0	9.5

Source: Medical Economics Company HMO/PPO Directory, 1993 and 1998.

Table 3. Managed Care Practices by Organization Type.

	Patient Selects Primary Care Provider		Capitated Payment	Discounted-FFS Payment
	1993	1998	1998	1998
HMO	48.9%	58.9%	53.8%	44.0%
Group/Staff	50.5%	56.3%	42.2%	14.1%
IPA/Network	50.0%	64.8%	60.0%	45.1%
Other	0%	11.3%	12.9%	8.1%
PPO	27.8%	34.2%	29.1%	50.6%
IPA/Network	33.6%	39.0%	33.1%	51.8%
Other	13.4%	17.1%	14.6%	23.6%
HMO-PPO	$X^2 = 37.792$ $p < 0.000$	$X^2 = 73.543$ $p < 0.000$	$X^2 = 75.9$ $p < 0.000$	$X^2 = 5.3$ $p < 0.01$

Source: Medical Economics Company HMO/PPO Directory, 1993 and 1998.

distribution of MC forms across regions of the country. The greatest decline in HMO prevalence occurred in regions where penetration was previously high, and the highest growth in PPOs occurred in those regions where HMOs were previously predominant.

Next we compare specific managed care practices in the different organizational forms. Table 3 shows that, overall, both HMOs and PPOs experienced growth in the use of primary care providers although HMOs are much more likely than PPOs to require patients to select a primary care provider (comparing the shaded rows of Table 3). This is only one measure of gatekeeping, in which the primary care provider is the source for referral to any specialty care. Managed care organizations may also require approval for referrals even when patients see a variety of providers – a practice most common in Group/Staff HMOs. Hence, the difference between HMOs and PPOs may be more attenuated if we had a better measure of gatekeeping practices.

Overall, HMOs are more likely to pay physicians through capitation, while PPOs are more likely to use discounted-fee-for-service to pay providers (compare shaded rows of columns three and four in Table 3). The HMO-IPA/Network model appears to be a hybrid form of MC by using both payment forms. Similar to PPOs, they have a relatively high use of discounted-fee-for-service, but like traditional HMOs, they also have a high use of capitation. These similarities and differences make sense if we consider, as others have (Sleeper et al., 1998), that health care provider organizations fall along a continuum, with traditional fee-for-service (indemnity-insurance) delivery at one end, and traditional staff HMOs at the other,

Table 4. Provider Oversight in Managed Care Organizations in the 1990s.

	Case Management		Second Opinion		Utilization Review	
	1993	1998	1993	1998	1993	1998
HMO	52.8%	56.7%	24.0%	36.5%	54.2%	57.3%
Group/Staff	52.7%	59.4%	23.7%	29.7%	51.6%	56.3%
IPA/Network	53.6%	61.9%	24.5%	40.9%	55.6%	62.9%
Other	25.0%	9.7%	8.3%	6.5%	25.0%	11.3%
PPO	42.1%	47.6%	24.5%	34.2%	47.2%	52.2%
IPA/Network	52.1%	55.5%	30.7%	40.6%	57.6%	60.7%
Other	17.5%	19.5%	9.3%	11.4%	21.6%	22.0%
HMO-PPO	$X^2 = 9.3$ $p < 0.001$	$X^2 = 10.1$ $p < 0.001$	$X^2 = 0.03$ n.s.	$X^2 = 0.071$ n.s.	$X^2 = 3.9$ $p < 0.05$	$X^2 = 3.2$ $p < 0.05$

Source: Medical Economics Company HMO/PPO Directory, 1993 and 1998.

with IPA/Network HMOs and PPOs in the middle. These hybrid forms account for some of the variation in managed care practices across markets, as the institutional pressures associated with operating as a hybrid form may be conflicting.

In Table 4, we compare the prevalence of provider oversight practices across years and organizational forms. Overall, provider oversight practices have increased in prevalence between 1993 and 1998. However, HMOs are more likely to use case management, somewhat more likely to do utilization review, and no more likely to require second opinions for surgery than PPOs (compare shaded rows in Table 4). Looking at the variation within models again reveals a slightly more complex picture. The IPA/Network PPOs look similar to various HMO models, but the newer, unspecified types (HMO-Other and PPO-Other) are much less likely to use provider oversight and case-management practices.

In Table 5, we compare general organizational characteristics of the various managed care forms. Overall, PPOs are more likely than HMOs to be for-profit firms. This finding stems primarily from the fact that only a little more than one-third of Group/Staff HMOs are for-profit firms. In contrast, HMO-IPA/Network models are nearly as likely as PPOs to be for-profit.

Overall, PPOs are much less likely to be nationally accredited or to issue consumer "report cards" with information about quality measures (compare shaded rows in columns two and three in Table 6). The variation on these two measures holds across model types and across years. Nearly 50% or more of the two dominant HMO models (Group/Staff and IPA/Network) are nationally accredited in both 1993 and 1998, compared to only about 13% of IPA-Network PPOs in each of those years. Similarly, approximately two-thirds of all HMO

Table 5. Organizational Features of Managed Care Organizations.

	% For-Profit		% Accredited		% Issue Report Card
	1993	1998	1993	1998	1998
HMO	65.2	64.8	60.1	55.5	67.0%
Group/Staff	45.1	37.1	66.7	57.8	63.8%
IPA/Network	69.0	68.2	55.6	49.4	66.8%
Other	90.0	64.5	0	19.4	77.3%
PPO	66.5	75.9	18.4	17.5	36.0%
IPA/Network	67.4	77.6	13.9	12.8	36.2%
Other	64.2	67.2	5.2	5.7	34.8%
HMO-PPO	$X^2 = 0.150$ n.s.	$X^2 = 15.9$ $p < 0.000$	$X^2 = 101.4$ $p < 0.000$	$X^2 = 130.2$ $p < 0.000$	$X^2 = 87.6$ $p < 0.000$

Source: Medical Economics Company HMO/PPO Directory, 1993 and 1998.

model types issue a quality report card, compared to only about one-third of PPOs.

What Do We Mean by Managed Care Anyway?

MC introduced the management of health care delivery primarily along two dimensions. One dimension introduced alternative methods for the payment of health services, moving from fee-for-service payment to mechanisms in which providers share risk (Luft, 1999). Some of the earliest HMOs paid providers by salary. More typically, MC organizations used capitation, as well as various forms of bonuses or penalties tied to utilization rates. Today, some plans now include discounted fee-for-service as a form of MC payment. The second dimension of MC introduced oversight of the clinical decisions of providers. Clinical management has included practices such as gatekeeping, utilization review, case management, and required second opinions. It is also possible to think of a third dimension by which MC "manages" delivery of health services, that of managing patient care-seeking by introducing patient co-payments for services, but we do not address this dimension here.

In the section above, we presented descriptive statistics of individual MC practices in different organizational forms. Here we present composite variables from the 1998 data measuring two dimensions of MC, payment mechanisms and clinical management, to assess the extent to which different organizational forms of MC actually "manage" health care delivery. Our first composite variable measures the clinical dimension of MC and includes whether an organization uses any of the following four practices of clinical oversight: case management, utilization review, second opinions or gatekeeping.

Table 6. Clinical Managed Care Practices in MC Organizations, 1998.

	Mean Number of Clinical MC Practices (0–4)	% with 2 + Clinical MC Practices
HMO	2.1	58.1
Group/Staff	2.0	59.4
IPA/Network	2.3	63.6
Other	0.4[a]	9.7
PPO	1.7	49.6
IPA/Network	1.96[b]	58.0
Other	0.7[a]	19.5
HMO-PPO	$F = 19.3$ $p < 0.000$	$X^2 = 8.9$ $p < 0.002$
Model Types	$F = 43.0$ $p < 0.000$	$X^2 = 131.3$ $p < 0.000$

Source: Medical Economics Company HMO/PPO Directory, 1998.
[a] Significantly less than all other model types ($p < 0.05$).
[b] Significantly less than Group/Staff-HMOs and IPA/Network-HMOs ($p < 0.05$).

Table 6 shows that, in general, HMOs use more clinical oversight practices to manage care than PPOs (compare shaded rows). A post-hoc Sheffe test reveals that this comparison holds even when comparing HMO-IPA/Network models with IPA/Network-PPOs (compare unshaded rows). Unspecified model types (HMO-Other and PPO-Other) use significantly fewer clinical MC practices than all others.

We analyze different forms of payment management, measured by two different composites, in Table 7. The first composite variable includes only the standard MC payment mechanisms of salary and capitation. The second composite includes discounted-fee-for-service, as well as salary and capitation. Well over half of all HMOs use either salary or capitation to pay providers, compared to less than one-third of all PPOs (compare shaded rows in column 3 of Table 7). IPA/Network-HMOs are more likely to use salary or capitation than either type of PPO, but are less likely than Group/Staff HMOs to use them. When we include discounted fee-for-service as a form of MC payment, the percent of PPOs with no MC payment mechanism declines, but PPOs are still less likely than HMOs to use payment management (see column 2 in Table 7).

Overall, HMOs have significantly more MC practices, including both clinical oversight and payment mechanisms, than PPOs. Both Group/Staff and IPA/Network HMOs have significantly more managed care practices than IPA/Network PPOs. On a continuum of "managedness," IPA/Network PPOs appear to fall between HMOs, on the one side, and traditional fee-for-service on the other. What is unclear, however, is whether and in which direction further change will

Table 7. Payment Management Practices in MC Organizations, 1998.

	Payment Mgmt 1: Salary or Capitation		Payment Mgmt 2: Salary, Capitation or DFFS	
	Mean	% No Payment Mgmt	Mean	% No Payment Mgmt
HMO	0.65	41.4	1.0	33.9
Group/Staff	0.91[a]	25.0	1.1[a]	25.0
IPA/Network	0.68[b]	38.1	1.1[a]	29.4
Other	0.13	87.1	0.21	82.3
PPO	0.33	69.3	0.78	42.8
IPA/Network	0.37	65.8	0.88	36.5
Other	0.20	82.1	0.43	65.0
HMO-PPO	$F = 99.2$ $p < 0.000$	$X^2 = 94.8$ $p < 0.000$	$F = 26.7$ $p < 0.000$	$X^2 = 10.0$ $p < 0.001$
Model Types	$F = 47.1$ $p < 0.000$	$X^2 = 166.6$ $p < 0.000$	$F = 32.9$ $p < 0.000$	$X^2 = 111.4$ $p < 0.000$

Source: Medical Economics Company HMO/PPO Directory, 1998.

[a] Significantly more than all other types ($p < 0.05$).

[b] Significantly less than Group/Staff-HMOs, but significantly more than other types ($p < 0.05$).

occur. Increasing numbers of PPOs suggest that managed care is becoming less managed over time.

DISCUSSION

The increasing prevalence of managed care has changed health care dramatically, including restructuring payments for health care services, reorganizing providers and services in health care markets, and introducing new practices that shape the relationships between and among health care providers and patients. We show that managed care currently consists of several distinct organizational forms that differ dramatically in their use of MC practices. PPOs use fewer MC practices and offer greater autonomy for providers than HMOs. PPOs also are less likely to be accredited and less likely to issue report cards. Some of these differences may exist because PPOs are a newer organizational form than HMOs. These differences, however, indicate that PPOs have distinctly different relationships to key stakeholders in the health care system than do HMOs.

The evidence also shows, however, that a substantial number of what are considered traditional HMOs do not use "traditional" MC practices, including utilization review and case management. This raises the question, are PPOs and

HMOs more alike or different? Future research needs to look more closely at local markets to determine whether competition among the different forms of MC occurs within local markets or whether local markets are dominated clearly by one form or another. Institutional theory suggests that competition among different forms in the local market is the major way in which institutional logics are questioned (Ingram & Clay, 2000; Knight, 1992). Subsequently, we must consider how competition among HMOs and PPOs is defining the future of managed care.

Those markets where HMOs and PPOs go into direct competition for consumers will be in turmoil institutionally. We may expect that within these markets, both types of organizational forms will have some change in their use of MC practices, as the social norms within the market are not clearly defined. In contrast, if PPOs have spread into markets where HMOs were never strong competitors, it would suggest that the strong managed care practices of HMOs were never fully accepted by the population and there will be little incentive (or institutional pressure) for either HMOs or PPOs to use strong MC practices. MC organizations may be highly susceptible to local variation for a number of reasons: because the relatively recent institutional development of its "logic" makes it more susceptible to regulatory pressure, because health care plans are "tightly coupled" with aspects of the local environment (e.g. supply of physicians, community employment level), or because norms within MC conflict with "traditional" values and practices of (fee-for-service) medicine.

CONCLUSION

A closer examination of the changes in organizational practices used across MC forms will help us gain a better understanding of how resource demands made by external stakeholders shape institutional processes. Institutional theorists have argued recently that bargaining among parties is a critical component of institutional change (Ingram & Clay, 2000). Many of the facets of patient-provider and provider-provider relationships are affected by the practices of managed care firms. For example, physicians' professional networks are often disrupted by MC rules (Anthony, 2003). While MC practices clearly affect consumers and medical professionals, we have little understanding of how these groups respond to changes in managed care practices – for example, by switching insurance as a consumer or switching practice locations as a physician (cf. Jiang & Begun, 2002). Such analyses require longitudinal data on how individual organizations change practices and lose or gain memberships. This study is part of a larger project to collect the organizational data necessary for addressing these questions.

The policy implications of the findings presented here depend more generally on how one views managed care. Some may conclude that the increasing number of PPOs is positive for consumers since PPOs entail less risk-sharing by providers and less gatekeeping of patients. In addition, PPOs typically offer a larger number of providers, thereby offering increased consumer choice and flexibility. Some will also believe that less clinical oversight more typical in PPOs is beneficial for both patients and providers. Others, however, may be worried that PPOs are throwing the managed care baby out with the bathwater. They may caution that the findings that PPOs are more likely to be for-profit, less likely to be nationally accredited, and less likely to issue report cards could signal quality differences between PPOs and HMOs. Moreover, it is not clear that PPOs yield the same cost savings as HMOs. While we presented no data on cost-differences, other studies have found that PPOs do not have the same savings as HMOs (Smith, 1997).

The data presented here illustrate how the organizational, forms and practices of managed care have changed in the last decade. Our study provides clues to some of the institutional processes that may be driving the industry to change and at the same time, demonstrates growing diversity and complexity in the organization of managed care. The diversity of practices both within and across organizational forms reveals a complexity that is often obscured by discussions and treatment of MC as a unitary entity. Social scientists attuned to the multidimensional features of institutional and organizational change can determine when, where and why the MC industry is changing in order to better understand the profound impact of MC on the delivery of health care.

NOTE

1. At least part of this difference can be explained by the different techniques for estimating enrollment between HMOs and PPOs. Unlike HMOs, in which "covered lives" equals the number of enrolled members, PPOs estimate the number of covered lives based on assumptions about the number of dependents per subscriber, and this "dependent factor" varies widely across plans (Smith & Scanlon, 2001).

ACKNOWLEDGMENTS

The authors appreciate comments from John L. Campbell and Ann Barry Flood, and helpful suggestions from the volume editor, Jennie Jacobs Kronenfeld. The research reported here was partly completed while Dr. Anthony was a Robert Wood Johnson Fellow in Health Policy at the University of Michigan. This research was also supported in part by a Rockefeller Social Science grant to the first author from the Rockefeller Center, Dartmouth College. An earlier version of this paper was presented at the ASA Annual Meetings, August 2000, Washington, D.C. Direct correspondence to Denise Anthony, Department of Sociology, HB 6104, Dartmouth College, Hanover, NH 03755. E-mail: Denise.Anthony@Dartmouth.edu

REFERENCES

Anthony, D. (2003). Changing the nature of physician referral relationships in the U.S.: The impact of managed care. *Social Science and Medicine*, *56*(10), 2033–2044.

Biggart, N. W., & Guillen, M. F. (1999). Developing difference: Social organization and the rise of the auto industries in South Korea, Taiwan, Spain and Argentina. *American Sociological Review*, *64*(5), 722.

Campbell, J. (1997). Mechanisms of evolutionary change in economic governance: Interaction, interpretation, and bricolage. In: L. Magnusson & J. Ottosson (Eds), *Evolutionary Economics and Path Dependence* (pp. 10–31). Cheltenham, UK: Edward Elgar.

Christianson, J. B., Sanchez, S. M., Wholey, D. R., & Shadle, M. (1991). The HMO industry: Evolution in population demographics and market structures. *Medical Care Review*, *48*, 3–46.

Dacin, M. T. (1997). Isomorphism in context: The power and prescription of institutional norms. *Academy of Management Journal*, *40*(1), 46–81.

Dalton, J. (1987). HMOs and PPOs: Similarities and differences. *Topics in Health Care Finance*, *13*(3), 8–18.

DiMaggio, P. J., & Powell, W. W. (1983). The iron cage revisited: Institutional isomorphism and collective rationality in organizational fields. *American Sociological Review*, *48*, 147–160.

DiMaggio, P. J., & Powell, W. W. (1991). Introduction. In: W. Powell & P. DiMaggio (Eds), *The New Institutionalism in Organizational Analysis* (pp. 1–38). Chicago: University of Chicago Press.

Dowell, G., & Swaminathan, A. (2000). Racing and back-pedalling into the future: New product introduction and organizational mortality in the U.S. bicycle industry, 1890–1918. *Organization Studies*, *21*, 405–431.

Edelman, L. B. (1992). Legal ambiguity and symbolic structures: Organizational mediation of civil rights law. *American Journal of Sociology*, *97*, 1531–1576.

Fligstein, N. (1990). *The transformation of corporate control*. Cambridge, MA: Harvard University Press.

Friedland, R., & Alford, R. R. (1991). Bringing back in: Symbols, practices, and institutional contradictions. In: W. W. Powell & P. J. DiMaggio (Eds), *The New Institutionalism in Organizational Analysis* (pp. 232–263). Chicago: University of Chicago Press.

Gabel, J. (1999). Job-based health insurance, 1977–1998: The accidental system under scrutiny. *Health Affairs*, *18*(6), 62–74.

Gabel, J., Liston, J., Jensen, G., & Marsteller, G. (1994). The health insurance picture in 1993: Some rare good news. *Health Affairs*, *13*(1), 327–336.

Gold, M., & Hurley, R. (1997). The role of managed care "products" in managed care "plans". *Inquiry*, *34*, 29–37.

Gold, M., Hurley, R., Lake, T., Ensor, T., & Berenson, R. (1995). A national survey of the arrangements managed-care plans make with physicians. *New England Journal of Medicine*, *333*(25), 1678–1683.

Greve, H., & Taylor, A. (2000). Innovations as catalysts for organizational change: Shifts in organizational cognition and search. *Administrative Science Quarterly*, *45*, 54–80.

Hannan, M., Carroll, G., Dundon, E. A., & Torres, J. C. (1995). Organizational evolution in a multinational context: Entries of automobile manufacturers in Belgium, Britain, France, Germany and Italy. *American Sociological Review*, *60*, 509–528.

Hannon, M., & Freeman, J. (1989). *Organizational ecology*. Cambridge, MA: Harvard University Press.

Haveman, H., & Rao, H. (1997). Structuring a theory of moral sentiments: Institutional and organizational coevolution in the early thrift industry. *American Journal of Sociology*, *102*, 1606–1651.

Hellinger, F. (1995). Selection bias in HMOs and PPOs: A review of the evidence. *Inquiry*, *32*, 135–142.

Ingram, P., & Clay, K. (2000). The choice-within-constraints new institutionalism and implications for sociology. *Annual Review of Sociology*, *26*, 525–546.

Jepperson, R. (1991). Institutions, institutional effects and institutionalism. In: W. Powell & P. DiMaggio (Eds), *The New Institutionalism in Organizational Analysis* (pp. 143–163). Chicago: University of Chicago Press.

Jiang, H., & Begun, J. (2002). Dynamics of change in local physician supply: An ecological perspective. *Social Science and Medicine*, *54*, 1525–1541.

Kitchener, M. K., & Harrington, C. (2003). Long-term care in the United States: A dialectic analysis of institutional dynamics. Working Paper.

Knight, J. (1992). *Institutions and social conflict*. New York: Cambridge University Press.

Light, D. (2001). Comparative institutional response to economic policy, managed competition and governmentality. *Social Science and Medicine*, *52*, 1151–1166.

Lounsbury, M. (2001). Institutional sources of practice variation: Staffing college and university recycling programs. *Administrative Science Quarterly*, *46*, 29–56.

Luft, H. (1999). Why are physicians so upset about managed care? *Journal of Health Policy, Politics and Law*, *24*(5), 957–966.

Mechanic, D. (1996). Changing medical organization and the erosion of trust. *Milbank Quarterly*, *74*(2), 171–189.

Mechanic, D. (2002). Socio-cultural implications of changing organizational technologies in the provision of care. *Social Science and Medicine*, *54*, 459–467.

Meyer, J., & Rowan, B. (1977). Institutionalized organizations: Formal structure as myth and ceremony. *American Journal of Sociology*, *83*, 340–363.

Miller, R. H., & Luft, H. S. (1994). Managed care plan performance since 1980. *Journal of American Medical Association*, *271*(19), 1512–1519.

Nelson, R. R., & Winter, S. G. (1982). *An evolutionary theory of economic change*. Cambridge: The Belknap Press of Harvard University Press.

Oliver, C. (1992). The antecedents of deinstitutionalization. *Organization Studies*, *13*(4), 563–588.

Ruef, M. (2000). The emergence of organizational forms: A community ecology approach. *American Journal of Sociology*, *106*(3), 658–714.

Ruef, M., & Scott, R. W. (1998). A multidimensional model of organizational legitimacy: Hospital survival in changing institutional environments. *Administrative Science Quarterly*, *43*, 877–904.

Scott, R. W. (2001). *Institutions and organizations* (2nd ed.). Thousand Oaks, CA: Sage.

Scott, R. W., Ruef, M., Mendel, P. J., & Caronna, C. A. (2000). *Institutional change and healthcare organizations: From professional dominance to managed care*. Chicago: University of Chicago Press.

Sleeper, S., Wholey, D. R., & Hamer, R. (1998). Trust me: Technical and institutional determinants of health maintenance organizations shifting risk to physicians. *Journal of Health and Social Behavior*, *39*(3), 189–200.

Smith, D. (1997–1998). The effects of preferred provider organizations on health care use and costs. *Inquiry*, *34*, 278–287.

Smith, D., & Scanlon, D. (2001). Covered lives in PPOs. *Medical Care Research and Review*, *58*(1), 16–33.

Stinchcombe, A. (1965). Social structure and organizations. In: J. G. March (Ed.), *Handbook of Organizations* (pp. 142–193). Chicago: Rand McNally.

Strang, D. (1995). Health maintenance organizations. In: G. R. Carroll & M. T. Hannan (Eds), *Organizations in Industry: Strategy, Structure and Selection* (pp. 163–182). Oxford: Oxford University Press.

Strang, D., & Bradburn, E. (2001). Theorizing legitimacy or legitimating theory? Neoliberal discourse and HMO policy 1970–1989. In: J. L. Campbell & O. K. Pedersen (Eds), *The Rise of Neoliberalism and Institutional Analysis* (pp. 129–158). Princeton: Princeton University Press.

Strang, D., & Soule, S. (1998). Diffusion in organizations and social movements: From hybrid corn to poison pills. *Annual Review of Sociology*, *24*, 265–290.

Suchman, M. C. (1995). Managing legitimacy: Strategic and institutional approaches. *Academy of Management Review*, *20*(3), 571–610.

Wade, J. B., Swaminathan, A., & Saxon, M. S. (1998). Normative and resource flow consequences of local regulations in the American brewing industry, 1845–1918. *Administrative Science Quarterly*, *43*, 905–935.

Wholey, D. R., & Burns, L. R. (1993). Organizational transitions: Form changes by health maintenance organizations. In: S. Bacharach (Ed.), *Research in the Sociology of Organizations* (Vol. 11, pp. 257–293). Greenwich, CT: JAI Press.

Wholey, D. R., Christianson, J. B., & Sanchez, S. M. (1992). Organization size and failure among health maintenance organizations. *American Sociological Review*, *57*, 829–842.

Wholey, D., Feldman, R., & Christianson, J. B. (1995). The effect of market structure on HMO premiums. *Journal of Health Economics*, *14*(1), 81–105.

Zucker, L. G. (1977). The role of institutionalization in cultural persistence. *American Sociological Review*, *42*, 726–743.

EDUCATION, MANAGED HEALTH CARE EXPERIENCES, AND HEALTH OUTCOMES

James W. Grimm, Zachary W. Brewster and D. Clayton Smith

ABSTRACT

Community household survey data tested the intervening role (between education and reported health outcomes) of adaptations of Antonovsky's (1987) tripartite sense of coherence (SOC). Comprehensibility was indexed by clarity and responsiveness of insurance representatives, manageability was measured by problems reported with physician office visits, and meaningfulness was assessed with household members' community health activities. SOC measures did not link education to either impairments or to health lifestyle scores. Comprehensibility and manageability linked education with self-reported well-being. Education and manageability each reduced impairments, while education, manageability, and meaningfulness increased lifestyle totals. Results help elucidate the influence of education on health.

Reorganizing Health Care Delivery Systems: Problems of Managed Care and Other Models of Health Care Delivery
Research in the Sociology of Health Care, Volume 21, 39–61

ISSN: 0275-4959/doi:10.1016/S0275-4959(03)21003-6

INTRODUCTION

The pattern of increasing morbidity and mortality as SES declines extends across most diseases, especially with respect to major causes of death such as heart disease (Cockerham, 1997; Link & Phelan, 1995, 2000; Marmot, 1996; Mirowsky, Ross & Reynolds, 2000; Robert & House, 2000; Wilkinson, 1996; Winkleby et al., 1992). Economic deprivation, higher risk social environments, and unhealthful behaviors (e.g. smoking) differentiate the higher rates of disease and death among the poor in comparison to the non-poor (Mirowsky & Hu, 1996; Winkleby et al., 1992). Public health insurance programs for the poor in the U.S. such as Medicaid have not offset the deleterious effects upon the health of the poor of economic deprivation and higher risk social environments (Dutton, 1986; Ross & Mirowsky, 2000).

Variation in health among the non-poor in the U.S. has not yet been explained adequately (Cockerham, 2001, p. 62). Educational attainment is the best single predictor of good health, and the differences in health by education level increase with age (Arber, 1997; House et al., 1994; Pappas et al., 1993; Ross & Wu, 1996). Previous research has demonstrated that educational effects upon health extend beyond the increased economic resources of the better educated (Mirowsky & Ross, 1998; Ross & Wu, 1996). Researchers have established that education fosters general orientations of personal control toward and management of the problems of daily living (including coping with health-related problems) that appear to sustain healthful living and good health (Feldman et al., 1989; Guralnik et al., 1993; Mirowsky & Ross, 1998; Pincus et al., 1998; Reynolds & Ross, 1998; Winkleby et al., 1992). These general orientations and the healthful behaviors they produce daily are of primary concern in the current research.

Various measures of the generalized sense of personal control over life events have been used to explain why, beyond their impact upon economic resources, the effects of education on health are so powerful. Two frequently used measures of the generalized sense of personal control in health-related research are Mirowsky and Ross' personal control (PC) (Mirowsky & Ross, 1998; Ross & Bird, 1994; Ross & Mirowsky, 1989, 1992; Ross & Wu, 1995) and Antonovsky's sense of coherence (SOC) (Antonovsky, 1979, 1987, 1993; Antonovsky & Sagy, 1986; Coe, 1997; Gallagher et al., 1994; Kivimäki et al., 2000). The former measure combines items from Rotter's (1966) locus-of-control scale and Pearlin et al.'s (1981) mastery scale, and includes such Likert-type items as "What happens to me mostly depends on me." and "I have little control over the things that happen to me." The latter measure includes items similar to those in Kobasa's (1982) "hardy personality" index and Bandura's (1977) "self-efficacy" measure and includes Likert-type indicators like "Do you have the feeling that you don't really care about what goes on around you?" and "How often do you have mixed up feelings and ideas?"

The PC measure has been used to explain the life orientations that link education to healthful living and to good health (Mirowsky & Ross, 1998). The SOC measure has been used to explain differences in coping and role overload among family caregivers (Coe, 1997; Gallagher et al., 1994). The general logic underlying both usages has been that people who feel that their daily lives are controllable, manageable, and meaningful will live more healthfully and continue to be in better health over their life course despite daily challenges and hassles (Antonovsky, 1987; Coe, 1997; Gallagher et al., 1994; Mirowsky & Ross, 1998; Pugliesi, 1995; Ross & Wu, 1995). While previous researchers have recognized that health experiences can reinforce and sustain general life events orientations, most of the causal interpretations have focused upon how such experiences shape the generalized orientations that result in better health and more healthful living. Experiences fostering a healthful orientation include factors such as challenging and autonomous employment (Pugliesi, 1995; Ross & Wu, 1995) and class of origin effects such as parents' education and values (Mirowsky & Ross, 1998).

Without denying that control orientations can and do influence healthful living and retaining good health, we believe that it is increasingly plausible to postulate that how members of households' perceive daily experiences in dealing with members' health problems shape their orientations toward health and their physical as well as emotional well-being. Households with fewer health problems, or those better able to mobilize the economic and caregiving resources to deal with the health problems that do exist, have the experiences that sustain the orientations that health is controllable and worth controlling. In studies using the SOC, for example, SOC levels were related to reports of role overload among family members caring for dementing patients but not those caring for patients with other conditions (Gallagher et al., 1994). In addition, parental caregivers to severely disabled children also had very low SOC (Margalit, Leyser & Avraham, 1989). Though PC has generally been related to healthfulness, the relationship may also have been conditioned by age and amount of physical impairments (Mirowsky & Ross, 1998).

BACKGROUND

We believe that a social constructionist view of household healthcare experiences is a viable way to consider health orientations and health-related experiences as reflexively related (Brown, 1995; Lorber, 1997). The social constructionist view of illness has been used to explain the diverse interpretations given by patients, family caregivers, and physicians to illnesses with essentially the same symptoms (such infectious diseases as pneumonia and gonorrhea) (Brown, 1995;

Lorber, 1997). Divergent interpretive meanings emerge over the course of diagnosing and treating illnesses from the broader social contexts of values, risk assessments, and social character (e.g. some illnesses are stigmatized while others are not). Such interpretive meanings both influence and reflect the manner in which chronic health problems are managed and judgments about the need for manageability (Bosk & Frader, 1990).

We extend the social constructionist viewpoint to include the interpretive meanings given to households' experiences in confronting members' health problems and their efforts to mobilize the resources to deal with these problems. The emergent interpretations of health-related resources reflect whether the actors believe the resources are sufficient to cope with health needs (Lin, 2001). For both resource-related and ideational reasons, the interpretations of the success or failure in mobilizing sufficient family resources to deal with health problems may be related to peoples' health and to their orientations toward health maintenance and control. In particular, negative interpretations may be related to both declining health and increased stress (Donelan et al., 1996).

To test our social constructionist perspective regarding orientations toward household health care experiences, we had to establish a context within which to compare health orientations and experiences. We employed a consumerism approach to needs deciding to measure the extent to which households needed a fuller range of health services beyond those of physicians. This broader range of needs is an improvement upon the use of physicians alone when considering indices of consumer needs (Chumbler & Grimm, 2000; Sofaer, 1998; Unland, 1998). We included among these households' health needs the services alternative providers proffer (such as chiropractors and acupuncturists, home health care professionals of all sorts, rehabilitative services, family caregiving, and clergy with whom family members might speak with about health problems) (Jones, 1997; Pescosolido, 1992).

This broad range of needed services has been used both to define consumer needs and to suggest the extent to which households are willing to choose and pay for services themselves if necessary (Bartlett, 1997; Cooper, Henderson & Dietrich, 1998; Cooper, Laud & Dietrich, 1998; Jones, 1997; Sofaer, 1998; Wylie-Rosett et al., 1995). Increasingly this broader range of service needs is recognized as appropriate both by patients and providers; however, frequently these services are contingent upon so-called patient agent organizations, including employers, health insurance programs, provider organizations including provider networks that contract for discounted pricing, and agencies of government (Kronenfeld, 2001). Thus, while increasingly necessary for health care needs, the exigencies of managed care mean that the linkages between patients needing services and having providers to give them what they need are increasingly tenuous

(Kronenfeld, 2001). Particularly problematic are the recent movements of HMO's away from services for the elderly as well as the restrictions on seeing specialists (Furlong & Wilken, 2001).

We developed our orientational measures using the conceptual framework rather than the item content of the SOC. We chose Antonovsky's SOC measure because its conceptual dimensions were better represented in our dataset than were the dimensions of PC and also because the three dimensions of the SOC were relevant in measuring orientations toward household healthcare experiences. In adapting our measures from the SOC, we tried to avoid two major problems with the SOC (Geyer, 1997). First, we were careful to control for health needs before drawing conclusions about orientations toward household health experiences. Second, we did not include any emotion items to remove the conceptual and content overlap between some of the SOC items and the manifestations of generalized anxiety and depression (Geyer, 1997). Our orientational measures were created for this analysis and include reactions to managed health care experiences.

SOC is a generalized orientation that expresses the degree to which people feel that their daily life is predictable, that resources are personally available to meet the needs of daily living, and that commitment to dealing with the demands of daily living is worthy of investment and engagement (Antonovsky, 1987, p. 19). The first component of the general orientation, *comprehensibility*, refers to the idea that people make cognitive sense of their daily experiences. *Manageability*, the second component, refers to the extent to which people perceive that resources are available to meet the challenges of their daily living. *Meaningfulness*, as the third component, refers to the extent to which people feel that their daily lives make sense concerning daily challenges being worth confronting as successfully as is possible (Antonovsky, 1987, pp. 16–19). In this study, the comprehensibility measure was developed with items showing satisfaction with the household's managed care plan and explanations of how it operated; the manageability measure was developed with items dealing with perceived problems with the household's managed care plan; and meaningfulness was indexed by a series of health-related behaviors that people in the household engaged in during the last year.

Comprehensibility

The comprehensibility of health care experiences in the era of managed care increasingly depends upon households being able to mobilize sufficient health-related resources in relation to household members' health needs (Kronenfeld, 2001; Lin, 2001). Neither family income (once above a level reflecting long-term economic deprivation) nor health insurance coverage are related to differences in health (Mirowsky & Hu, 1996; Ross & Mirowsky, 2000). Many people who report problems in paying for and obtaining health care in the last year are insured, and like the uninsured, they say that such problems exacerbated serious health problems (Donelan et al., 1996). Having adequate health insurance coverage increasingly depends upon exigencies of employment such as whether employers provide such benefits and, if they do, the extent to which coverage is sufficient to meet subscribers' needs (Seccombe & Amey, 1995). Increasingly, households must successfully negotiate a host of intermediary hurdles before the services of providers can be secured (Kronenfeld, 2001). Though patients have received increasing rights through COBRA legislation, health insurance becomes an out-of-pocket expense when people lose their jobs. In addition, the increasing number of subscribers in for profit HMOs has meant increased restrictions on services and limits on pharmaceuticals (Furlong & Wilken, 2001).

When households, rather than individual subscribers are studied, the limits of insurance coverage and the difficulties in seeing physician specialists and other needed providers become more problematic (Kronenfeld, 2001; Robinson, 1997). The emotional strain and difficulties in dealing with all the household problems at once are primary reasons why the effects of SOC scores upon stress and role overload have been situation-specific rather than general (Gallagher et al., 1994). In this more complex and problematic patient-physician-other provider's environment, increasingly states have legislated mechanisms for dealing with patient complaints (Kronenfeld, 2001). The items used to measure the comprehensibility of household health experiences in the last year include evaluative reactions to the services of the plan and how well the services were explained when members of the household had questions about the plan and its services.

Manageability

The perceived manageability of household health problems depends upon many more complicating conditions in the managed care era than it did in the traditional fee-for-service system when the quality of the physician-patient relationship was of paramount importance (Kronenfeld, 2001). Now, the manageability of health care in relation to patient needs depends on multiparty patient agents like employers, health insurance management organizations, governmental regulations, and a variety of contractual stipulations in the annual adjustment to coverage limits, preferred provider networks, premiums, deductibles, and co-payments (Hafferty & Light, 1995; Kronenfeld, 2001; Sofaer, 1998). Thus, the interrelated health needs of household members are related in increasingly complicated ways to both household health resource mobilization (Lin, 2001) and to the exigencies of managed care causing attending to all members' health needs to be increasingly problematic (Kronenfeld, 2001). Recent studies have shown that women in particular may experience delays in and/or disruptions of treatment (Furlong & Wilken, 2001). Consequently, insured households encounter problems in arranging for and in paying for their required portions of healthcare, particularly when chronic health needs require expensive medications (Bodenheimer, 1996; Friedman, 1997; Vladeck, 1999). The range of items that we used to measure perceived manageability included a series of perceived problems that members of the households had in arranging and receiving healthcare, especially from physicians.

Meaningfulness

The meaningfulness of staying healthy is patterned by educational attainment (Mirowsky & Ross, 1998; Ross & Wu, 1995) and reflects the fact that well educated parents impart health maintenance values to their children (Mirowsky & Ross, 1998). Considerable evidence from prior studies has shown that motivation to control health and emotional well-being are related to the self-enhancing characteristics of the challenging and autonomous work better educated people do (Mirowsky & Ross, 1998; Pugliesi, 1995; Ross & Wu, 1995). To eliminate the effect of SOC indicators being related to the emotional manifestations of anxiety and depression (Geyer, 1997), and to focus our indicators upon commitments to and engagement in health enhancing activities per se (Antonovsky, 1987), our indicators of meaningfulness of health included a wide range of health-related activities that members of households engaged in during the last year such as attending wellness seminars, donating blood or plasma, getting preventive screens, donating to health related causes like the Red Cross, and volunteering in healthcare facilities like nursing homes and hospices. These types of health activities have been shown to link personal health values and community health related norms (Wellman & Frank, 2001).

HYPOTHESES

We consider the three orientational measures of households' health care experiences in relation to managed care to be important intermediate links between educational attainment and health-related output measures that include physical impairments, well-being, and health lifestyle. We expect the orientations to help explain differences in health outcomes that the influence of education alone cannot.

That is, education is linked to health by experiences that sustain the comprehensibility of health care, its manageability, and the continued commitment to behaviors that foster health. We test the following hypotheses concerning how we expect the orientational measures to intervene between education and health-related outcomes.

(1) Net of household contextual differences including health needs, and respondents' age and gender, better educated people whose household health experiences are perceived as more comprehensible and more manageable and for whom health activities are more meaningful, will have fewer physical impairments.
(2) Net of household contextual differences including health needs, and respondents' age and gender, better educated people whose household health experiences are perceived as more comprehensible and more manageable, and for whom health activities are more meaningful, will have higher well-being scores.
(3) Net of household contextual differences including health needs, and respondents' age and gender, better educated people whose household health experiences are perceived as more comprehensible and manageable, and for whom health activities are more meaningful, will have higher health lifestyle scores.

METHODS

Research Context

The data used in this study were collected from a randomly selected sample of households ($n = 182$) in Warren County, Kentucky. The county (population = 100,000) is dominated by Bowling Green (population = 50,000), which is a regional cultural and economic center and where Western Kentucky University (16,000 students) is located. The university is the largest employer in the county, and the university's influence is reflected by the fact that 46% of the respondents had completed at least four years of education. Most respondents (93%) were employed at least part-time, and two-thirds (64%) of the members of the sample reported annual family incomes of $30,000 or more. Most respondents (over three-fourths) were from twenty-five through sixty-four years of age, and only 7.1% were more than sixty-five. Fifty-eight percent of the members of the sample were females and 42% were males. Eighty-five percent of the Warren County Health Survey (WCHS) respondents were in households covered by managed healthcare and nine of ten of those households obtained such care through employers.

Sampling

A random digit dialed (RDD) technique was used by the Western Kentucky University Social Research Laboratory to select households in the 1998 Warren County Health Survey (WCHS). RDD procedures combined proportionate representation of user prefix areas and four randomized digits. Enough randomized numbers were generated to compensate for the nonresidential numbers and refusals and still obtain a desired sample size of about how two hundred. Randomized variation was used to select respondents within households and substitutions were not allowed. Up to six callbacks were made to reach desired respondents. The overall response rate (the percent of completed interviews among contacted eligible numbers) was 48%, as compared with the 42% response rate for surveys of the public conducted during 1997, using RDD techniques and with interviews lasting about twenty minutes (Council For Marketing and Opinion Research, 1998). The distributions of demographic characteristics in the WCHS sample were not significantly different from those in the 1990 U.S. Census data available for Warren County Kentucky, regarding age, being married, being white versus being nonwhite, and gender. The WCHS sample was representative of the study population.

Interviewing

Well-trained interviewers who were thoroughly familiar with the survey instrument and used standardized follow-up and explanatory guidelines conducted telephone interviews lasting an average of twenty minutes. Interviewers used extra time for those respondents who needed it to comprehend and respond to the questions. Interviews were completed between 6:30–8:30 p.m., Monday through Thursday, from mid-October through mid-November 1998. Care and skill in follow-ups results in negligible item non-response and the missing data was small enough that no substitution of values was necessary.

Measures

Several covariates in the present analyses were respondents' socio-demographic traits. These include educational attainment, measured with these five degree levels: less than high school = 1, high school or GED = 2, technical or junior college degrees = 3, baccalaureate degrees = 4, and post graduate degrees = 5. Age was measured as a continuous variable and gender was coded female = 1, and male = 0. Other covariates included household contextual variables including composition measured by the number of people living in the household and annual household income, which was measured with this eight-point scale: 1 = less than $10,000; 2 = $10–19,999; 3 = $20–29,999; 4 = $30–39,999; 5 = $40–49,999; 6 = $50–69,999; 7 = $70–89,999; and 8 = $90,000 or more.

The SOC indices we developed were based upon the original three dimensions of Antonovsky's SOC measure: comprehensibility, manageability, and meaningfulness (Antonovsky, 1987). All the emotional indicators in the original SOC were replaced with orientational and behavioral measures regarding household experiences with managed care in the last year. For comprehensibility we chose seven items showing how much satisfaction respondents expressed (very satisfied and satisfied = 1 versus not sure, dissatisfied or very dissatisfied = 0) on a Likert-type series of questions regarding respondents' satisfaction with their health care plan, service representatives of the plan, explanations of benefits, answers to health insurance questions, physicians' explanations of medications, the doctors in the plan, and the services of the entire plan. These items well represented the range of experiences related to subscribers' ability to understand and comprehend health care services (Kronenfeld, 2001) and the item on physicians represents Antonovsky's point that the comprehensibility of daily living is dependent upon the quality of relationships with supportive others, and for health with physicians or psychotherapists in particular (Antonovsky, 1987). The items in the comprehensibility scale had high internal reliability (alpha = 0.83), and comprehensibility scores ranged in value from one through seven with higher scores indicating increased comprehensibility.

The manageability scale with respect to households' health care was assessed through respondents' evaluations of household members' experiences in the last year with managed care. A ten-item series dealt with costs of visits to physicians, test results being returned promptly, amount of time physicians spent during office visits, ease in making appointments, waiting time in doctors' offices, experiencing coverage limits, and having to take a second job to meet health care costs. We feel that this range of items accurately represents the problems increasingly related to subscribers' satisfaction with managed care in relation to their health needs (Kronenfeld, 2001). Response patterns showed good internal consistency (alpha = 0.70) and scores ranged from one through ten. High scores reflected increased manageability and the absence of problems in experiencing health care.

Measuring the meaningfulness of good health also led us to omit emotion items and use behavioral indications of important health and health-promoting activity.

The items we include about members of the sampled households are those that have been found to link personal health values with support for community health norms (Wellman & Frank, 2001). We used eleven items indicating in the past year whether (=1) or not (=0) a member of the household had health screens (blood pressure checks, cholesterol checks, prostate tests, or mammograms), had general physicals, volunteered in a health-related program or organization, donated blood or plasma, taken herbal medicine or vitamins, and donated money to a health charity or to a healthcare facility. The meaningfulness items had good internal reliability (alpha = 0.72) and higher scores indicated increased meaningfulness. Scores ranged from zero through eleven.

Each SOC orientation component measurement was carefully developed on the basis of the face validity of the items available in the data set. Consistent with prior use of the original SOC, the three indices did not form a unidimensional whole (Antonovsky, 1987, 1993; Gallagher et al., 1994). The correlations among the three scale scores were moderate, indicating that they form a three-dimensional representation of SOC in relation to perceptions of households' managed health care experiences in the last year.

We include three self-reported health outcome variables in the present analysis: physical health, physical and emotional well-being, and health lifestyle. Physical health was measured using a subset of physical functioning items in the Medical Outcome's Study's 36-item short form (SF-36) (Ware & Sherbourne, 1992). Items included respondents' answers regarding how often, where 0 = never, 1 = a few days, 2 = some days, 3 = most days, and 4 = every day, in the last month their own health problems limited them in climbing stairs, bending or stooping, walking one block, walking several blocks, bathing or dressing, lifting or carrying groceries, moderate activity like moving a table or pushing a vacuum, heavy lifting or running, or performing normal work outside the home. This is a well-developed scale of physical impairments and the internal consistency of the WCHS respondents' answers was very high (alpha = 0.93). Since a little over half (56%) of them had no impairments and half of those who did have impairments had scores below seven, the index clearly measured moderate impairment.

Respondents' self-assessed indications of their general physical and emotional well-being was measured by a series of nine items concerning health status in the month before being interviewed. This measure consisted of Likert-type items on which respondents indicated how often (4 = everyday, 3 = most days, 2 = some days, 1 = very few days, 0 = never) they felt full of energy, not nervous, not exhausted, happy, calm and peaceful, not tired, not in severe pain, had no problems sleeping, and did not feel stressed. The items had considerable internal validity (alpha = 0.82) and scores could range from 4 through 36, with increasing values indicating better physical and emotional well-being.

Health lifestyle was measured with a two-factor index comprising a combination of both positive and negative health related behaviors: exercising one-half hour daily, sleeping well, taking time to relax, not smoking, and not driving above legal speed limits. The first factor related to respondents' positive engagement in daily habits of living (sleeping well, taking time to relax, and exercising one-half hour daily, where 4 = everyday, 3 = most days, 2 = some days, 1 = a few days, and 0 = never) in the month before interviewing. Using the same Likert measurement, the other factor indicated the absence of negative health-related behaviors (not smoking, and not driving above legal speed limits). The scores on this two-factor index ranged from 0 through 20. This measure reflects the combination of daily patterns of living that influence health maintenance.

Analyses

Multiple regression with tests for interaction, multicollinearity, and curvilinearity (Ganzach, 1998) were used to test hypotheses. Regression models were developed by regressing respondents' physical health, self-assessed indicators of general physical and emotional well-being, and health lifestyle scores on respondents' age, gender, and education, households' composition, income and health needs, and the adapted SOC measures. In each case, we entered the respondents' traits and household contexts first (Model 1) followed successively by each of the three orientations scales (Models 2–4). The effects of covariates, education, and the health orientations are assessed as each orientational measure is added by comparing the net effects of predictors and also by increments in R squared. We assess the extent to which the education effect is reduced by subsequent entries as a test of orientations intervening to link educational effects to health related outcomes. Results of the testing for interaction, multicollinearity, and curvilinearity were used to add a Model 5 for health lifestyle scores. Hypotheses are evaluated through examination of the direction and statistical significance of net effects (betas), the increments in R squared as components are added to the models, the interaction effects and nonlinear effects through additional net effects (betas), and the decline in education effects with subsequent consideration of orientation measures.

RESULTS

Regression results related to the variation in WCHS respondents' physical impairments are in Table 1. Excluding the orientation measures, in Model 1 impairment differences are contingent upon age (0.396) and education (−0.184). About 28%

Table 1. Regression of WCHS Respondents' Physical Impairment Scores on Covariates and New Adapted SOC Measures: Comprehensibility, Manageability, and Meaningfulness.

	Model 1	Model 2	Model 3	Model 4
Household health needs	0.073	0.070	0.026	0.003
Household income	−0.111	−0.105	−0.102	−0.108
Household composition	−0.100	−0.099	−0.103	−0.112
R's being female	0.082	0.086	0.088	0.082
R's age	0.396***	0.401***	0.400***	0.386***
R's education	−0.184*	−0.180*	−0.164*	−0.172*
Comprehensibility		−0.031	0.020	0.010
Manageability			−0.176	−0.172*
Meaningfulness				0.065
Manageability^2			NS	NS
Meaningfulness^2				NS
R^2	0.282***	0.283***	0.308***	0.311***

*$p < 0.05$.
***$p < 0.001$.

of the differences in impairment are explained by Model 1 ($R^2 = 0.282$). Adding the comprehensibility effect in Model 2 does little to alter the effects found in the first model, and comprehensibility itself did not affect impairments. The addition of manageability in Model 3 increased the percent of variation in impairments explained to nearly 31% ($R^2 = 0.308$). The effect of manageability upon impairments is negative (−0.176) and the effects of both age (0.400) and education (−0.164) remained essentially unchanged. The meaningfulness measure added in Model 4 had no significant impact upon impairments, and the effects of age (0.386), education (−0.172) and manageability (−0.172) remained about the same as in Model 3.

We found very little support for the first hypothesis. The effect of education in relation to fewer impairments was expected, but that effect was not diminished as it would have been if the predicted mediating influences of orientations had been found. Instead, results clearly showed that the most important influence in relation to impairments was the aging process. The net effect of educational attainment was to reduce impairments, but results showed that manageability of household health needs were just as strongly related to fewer impairments as educational attainment was (−0.172). Rather than mediating the influence of education upon physical health problems, the manageability of health needs was just as important as education in relation to fewer physical problems. While better educated people had fewer health problems, so did people who reported household health problems as

Table 2. Regression of WCHS Respondents' Well-Being Scores on Covariates and New Adapted SOC Measures: Comprehensibility, Manageability, and Meaningfulness.

	Model 1	Model 2	Model 3	Model 4
Household health needs	−0.138^{+}	−0.118^{+}	−0.084	−0.041
Household income	0.220**	0.169*	0.167*	0.175*
Household composition	−0.061	−0.073	−0.069	−0.052
R's being female	−0.087	−0.115	−0.118	−0.106
R's age	0.025	−0.016	−0.016	0.011
R's education	0.153*	0.116	0.104	0.120
Comprehensibility		0.260**	0.220**	0.239**
Manageability			0.137^{+}	−0.130^{+}
Meaningfulness				−0.119
Manageability^2			NS	NS
Meaningfulness^2				NS
R^2	0.120**	0.180***	0.196***	0.206***

$^{*}p < 0.05$.
$^{**}p < 0.01$.
$^{***}p < 0.001$.
$^{+}p < 0.10$.

manageable concerning the exigencies of their managed care plans. The effects of education and manageability on impairments were independent and additive and no interactive effects or exponential effects were found in relation to impairments.

Findings regarding the variation in WCHS respondents' well-being scores in Table 2 provided strong evidence in support of the hypothesized intermediary role of orientations linking education to better well-being. Sans orientational measures, results showed that variation in well-being was dependent upon household income (0.220) and education (0.153), as well as household health needs (−0.138). The first model explained 12% of the variation in well-being ($R^2 = 0.120$). Adding the comprehensibility measure in Model 2 substantially increased the amount of variance in well-being explained ($R^2 = 0.180$), and the effect of education was substantially attenuated. The addition of comprehensibility also substantially reduced the impact of both income (0.169) and needs (−0.118), making the effect of comprehensibility (0.260) the most important influence in the second model. The addition of manageability in Model 3 statistically attenuated the needs effect and slightly reduced the comprehensibility effect, but the income effect remained significant (0.167). The third model explained about 20% of the variation in well-being scores ($R^2 = 0.196$). Adding the meaningfulness effect in Model 4 slightly increased how much variation in well-being accounted for ($R^2 = 0.206$), and while the manageability effect was diminished some, the income and comprehensibility effects were increased, when compared with the third model.

We found clear support for the second hypothesis. The effect of education upon well-being scores was mediated by both the comprehensibility of household health care expenses and by the perceived success of the managed care plan in dealing with household health needs. The impact of household health needs upon well-being was also substantially mediated by comprehensibility and manageability. Household needs influence well-being largely through the process by which the exigencies of managed care in dealing with needs are understood and viewed as adequate for household health needs.

Results of the analysis of differences in the WCHS respondents' health lifestyle scores appear in Table 3. Not considering any orientational measures, health lifestyle scores were only dependent upon respondents' age (0.201) in Model 1. The addition of the comprehensibility effect in Model 2 showed that lifestyle scores were dependent upon comprehensibility (0.164) and the age effect was diminished slightly (0.173). The addition of the manageability effect in Model 3 more than doubled the proportion of explained differences in lifestyle scores ($R^2 = 0.166$). Manageability was strongly related to lifestyle (−0.995); in fact, the effect of manageability turned out to be exponential (1.268 betas) when we tested for

Table 3. Regression of WCHS Respondents' Health Lifestyle Scores on Covariates and New Adapted SOC Measures: Comprehensibility, Manageability, and Meaningfulness.

	Model 1	Model 2	Model 3	Model 4	Model 5
Household health needs	−0.082	−0.069	0.006	−0.007	−0.033
Household income	−0.053	−0.084	−0.065	−0.051	−0.031
Household composition	0.034	0.027	0.052	0.009	−0.019
R's being female	0.008	−0.011	0.012	0.019	0.010
R's age	0.201*	0.173*	0.155*	0.166*	0.153*
R's education	0.108	0.084	0.065	0.045	0.868**
Comprehensibility		0.164*	0.088	0.114	0.130
Manageability			−0.995*	−0.935*	−0.755*
Meaningfulness				−0.744*	−0.686*
Manageability^2			−1.268**	1.196**	1.597**
Meaningfulness^2				0.086**	0.757**
Manageability × Education					−1.092**
R^2	0.048	0.072	0.166**	0.208***	0.244***

*$p < 0.05$.
**$p < 0.01$.
***$p < 0.001$.

curvilinearity. A very small increase in manageability resulted in a very large decline in lifestyle scores. Moreover, adding the manageability effect further diminished the age effect (0.155). Adding meaningfulness in Model 4 showed that it, too (−0.744), was exponentially related to lifestyle scores (meaningfulness^2 = 0.806). The fourth model explained about 21% of the variation in lifestyle scores ($R^2 = 0.208$).

Testing for interaction yielded results that required a fifth model, when the interaction term between manageability and education (manageability × education) was added. The fifth model showed that orientations did not mediate the effect of education but with it conjointly influenced lifestyles. Furthermore, inclusion of the interaction term resulted in education having separate and opposing effects upon lifestyle scores, both of which were not mediated by orientations. On the one hand, the net effect of education on lifestyle scores is positive (0.868). On the other hand, education acting conjointly with manageability had a powerful effect in reducing lifestyle scores (−1.092 betas). Furthermore, the exponential effects of both manageability and meaningfulness meant that small increments in either effect resulted in substantial declines in lifestyle scores.

Results concerning the effects of orientations upon lifestyle scores do not support the third hypothesis. Orientations do not link education to higher lifestyle scores regarding the indicators of healthy living that we included in our analysis. Remember that some involved time-bound activities such as daily exercise, daily relaxation, and sleeping well while others involved avoiding risk behaviors such as smoking and driving in excess of speed limits. Net of its relationships with orientations, education increased health lifestyle scores. When combined with manageability, however, education was related to much lower scores. In addition, a small increase in manageability or meaningfulness lowered health lifestyle scores. Both our measures of manageability and meaningfulness involved time invested activities in taking care of household health needs and being involved in various community health-related programs. Consequently and ironically, health related activities and obligations appear to make the time bound activities of health lifestyles more problematic.

DISCUSSION

We do not believe that our results challenge nor do we intend them to be taken as a challenge of the important role of life orientations in relation to health and healthful living (Mirowsky & Ross, 1998; Pearlin et al., 1981; Pugliesi, 1995). The generalized personal control and greater self-efficiency generated by education and subsequent avoidance of economic deprivation and other problems in adulthood (Ross & Wu, 1995) are very important considerations in explaining why better-educated people are healthier. However, we feel that our results clearly show the utility of considering households' experiences with managed care plans in relation to members' health and habits of living (Donelan et al., 1996). While being insured is not related to health (Mirowsky & Ross, 1998), our results suggest that the appropriateness of the plan and the quality of its services vis-à-vis households health needs are now important considerations in relation to health (Kronenfeld, 2001). Because obtaining managed care is increasingly complex and problematic, the orientations of subscribers toward their plans and services provided by them may result in healthcare orientations being much more dynamic than the presumed stability of generalized control orientations (Antonovsky, 1987). Such dynamics may help explain why some recent studies have found weak longitudinal SOC effects (Kivimäki et al., 2000). Evidence regarding delays and disruptions of treatment, especially for women also may have significant effects on SOC (Furlong & Wilken, 2001).

We do not dismiss the importance of health beliefs for motivating people to control their health, to follow healthful habits of living, and to screen and seek treatment quickly. However, our results suggest two complications regarding the effects of health beliefs. First, we have found that the meaningfulness of good health may be less important in relation to being in good health than the resources for and the exigencies of health insurance plans to provide necessary health care. In particular, we have found that the manageability and the comprehensibility of household health experiences were much more related to well-being than was the meaningfulness of health. Second, we have found that people for whom health is more meaningful, and who may be better able to manage the health needs of their households, appear themselves to have less time available for those daily time-bound health lifestyle behaviors such as exercise and relaxing. Future research must address through longitudinal research the longer range implications for stress and health lifestyle of the increasing complex and problematic process of addressing necessary household health needs in the era of managed care (Kronenfeld, 2001).

Our results clearly support the previous researchers who have found that both educational and economic resources sustain healthfulness (Mirowsky & Ross, 1998; Ross & Wu, 1995). Employment and adequate economic resources help sustain the self-efficiency necessary for controlling health (Mirowsky & Ross, 1998; Pugliesi, 1995). Our results suggest, however, that healthcare orientations including the perceived adequacy of health insurance coverages and services not so much link education to health as they act as important independent forces that inhibit a continuation of being able to control health even when people understand how important that it is to do so. Our study has shown that when households had limits imposed by employer provided insurance and other problems in obtaining managed care services (three-fourths of the WCHS sample were covered by employer provided insurance) respondents reported significantly more impairments and lower well-being scores. Health problems also may be compounded as more workers, including elderly employees, find themselves working part-time or limited hours of employment, which traditionally do not carry sufficient health insurance benefits (Marshall, 2001).

Future research, especially longitudinal studies, also should pay more careful attention to age effects upon health. Our results clearly show that education, resources, and managed care orientations are not as important separately or in combination as is aging, with respect to increasing impairments. How better to address and deal with initial impairments seems necessary to forestall the multi-illnesses and chronicity of disabling conditions. Equally important, we feel, is the need for finding out why self-indications of well-being are not more related to age. Stress and emotional problems and daily hassles may be better handled by people with better developed sense of control (Antonovsky, 1987, 1993). However, the toll that stress and generalized anxiety takes on people who do not necessarily experience anything other than daily hassles may be quite severe. Yet, the focus of the previous research has been upon major stressful events like divorce or joblessness and has tended to ignore the daily struggle now required to manage daily living including daily household health needs. We suspect that important avenues for research will be the deteriorating employer-provided benefits (Marshall, 2001) and the increasing family care giving burden placed upon employed adults (Gallagher et al., 1994).

Clearly the limits of our cross-sectional data do not enable us to conclude either that insurance exigencies and managed care services effect people's health definitively or that more illness is why more problems with managed care occur. Finding that households in which managed care services have involved fewer problems are those in which respondents had fewer impairments and higher well-being scores is encouraging, however. Thus, while being insured is not related to health (Ross & Mirowsky, 2000), our results suggest that the adequately managed care services in relation to household health needs, may be related to less illness and less stress. Future research must attempt to evaluate such causality. On the other hand, it is important that future studies carefully investigate the real possibilities that problems in using managed care to deal with multiple illnesses among multiple subscribers does in fact diminish both health and well-being (Furlong & Wilken, 2001; Kronenfeld, 2001). At the very least, such evaluation studies would provide the necessary evidence for concluding how to improve and extend the managed services in both cost effective and treatment effective ways.

We believe that our results concerning the relationships between education, managed care orientations, and health lifestyles highlight a very important irony. Better educated people, for whom health is very important, are the very people who must confront and successfully deal with the complex exigencies of using managed care to address household health needs (Kronenfeld, 2001). The commitment and engagement now necessary to manage successfully household health precludes better educated people themselves from engaging in some forms of daily activity necessary for healthfulness. This seems particularly true of activities like exercise. Having the time and daily schedules to do so is very problematic. What our study demonstrates is that approaches to healthful living should address the ways in which busy well-educated people can accommodate and regularly pursue one-half hour of daily exercise and other time-bound activities that are more difficult to arrange. Our results thus provide a good reason why people who believe in healthfulness and who are otherwise relatively healthy still do not regularly exercise.

Our results have provided very mixed results concerning the linkages between education and health. We feel it is very important to caution that we do not feel that our results diminish nor do we intend them to diminish the importance of education for health. That we have found health orientations link education to some health outcomes but not others only heightens the need for a better understanding of the intervening experiences with managed care that we have found have independent and potentially disruptive effects between education and health outcomes. Moreover, because education heightens both the meaningfulness of healthfulness and commitment to coping with health needs, our results suggest it is important to understand why better educated people have less time to complete time-bound healthful habits in their daily lives. Future research must try to find answers for the behavioral dilemmas involved in arranging and obtaining health care and being healthy. We feel that it is very important that we found education as having positive net effects upon health, once the problems of engagement in obtaining managed care were removed. Education, net of the management of health care creating problems in maintaining daily health habits, had clear positive effects on impairments and health lifestyle. Consequently our study confirms the very important ways education impacts health beyond its role in mobilizing the resources to live healthfully.

CONCLUSION

Future research must better address how the exigencies of employer provided insurance relate to household health needs and serving them. Of particular concern in doing so are the ways in which so-called agent organizations intervene to complicate the management of health care services in households which involve well-educated people, especially older people, who may want to be healthy but encounter many problems in managing their health (Furlong & Wilken, 2001; Kronenfeld, 2001; Marshall, 2001). Despite its very important role in helping mobilize economic and other household resources for health care including family caregiving, it is time to address when and how education ceases to be linked to health outcomes. Knowing more about when and why educational effects are disrupted in relation to health will be a way to improve the health of all Americans.

ACKNOWLEDGMENTS

The authors wish to thank Neale Chumbler and Jennie Jacobs Kronenfeld for their assistance with this article.

REFERENCES

Antonovsky, A. (1979). *Health, stress, and coping*. San Francisco, CA: Jossey-Bass.

Antonovsky, A. (1987). *Unraveling the Mystery of Health: How people manage stress and stay well.* San Francisco, CA: Jossey-Bass.

Antonovsky, A. (1993). The structure and properties of the sense of coherence scale. *Social Science and Medicine, 36*(6), 725–733.

Antonovsky, A., & Sagy, S. (1986). The development of a sense of coherence and its impact on stress situations. *Journal of Social Psychology, 126*(2), 213–225.

Arber, S. (1997). Comparing inequalities in women's and men's health: Britain in the 1990s. *Social Science and Medicine, 44*(6), 773–787.

Bandura, A. (1977). Self-efficacy: Toward a unifying theory of behavioral change. *Psychological Review, 84*(2), 191–215.

Bartlett, D. F. (1997). Preparing for the coming consumer revolution in health care. *Journal of Health Care Finance, 23*(4), 33–39.

Bodenheimer, T. (1996). The HMO backlash – righteous or reactionary? *The New England Journal of Medicine, 335*(21), 1601–1604.

Bosk, C. L., & Frader, J. E. (1990). Aids and its impact on medical work: The culture and politics at the shop floor. *Milbank Quarterly, 68*(4), 257–279.

Brown, P. (1995). Naming and framing: The social construction of diagnosis and illness. *Journal of Health and Social Behavior* (Extra Issue), 34–52.

Chumbler, N. R., & Grimm, J. W. (2000). Channels of podiatrists' referral communication to physicians. *Journal of Applied Sociology, 17*(1), 69–85.

Cockerham, W. C. (1997). *This aging society* (2nd ed.). Upper Saddle River, NJ: Prentice-Hall.

Cockerham, W. C. (2001). *Medical sociology* (8th ed.). Upper Saddle River, NJ: Prentice-Hall.

Coe, R. M. (1997). The magic of science and the science of magic: An essay on the process of healing. *Journal of Health and Social Behavior, 38*(1), 1–8.

Cooper, R. A., Henderson, T., & Dietrich, C. L. (1998). Roles of non-physician clinicians as autonomous providers of patient care. *Journal of the American Medical Association, 280*(9), 795–801.

Cooper, R., Laud, P., & Dietrich, C. L. (1998). Current and projected workforce of non-physician clinicians. *Journal of the American Medical Association, 280*(9), 788–794.

Council For Marketing and Opinion Research (1998). Industry watch. Port Jefferson NCMOR. Reprinted by Permission (1999) in Industry Survey: CMOR's Respondent Cooperation Audit. In: *Sawtooth News: News from Sawtooth Technologies on Computer Interviewing and Analysis, 15*, p. 4.

Donelan, K., Blendon, R. J., Hill, C. A., Hoffman, C., Rowland, D., Frankel, M., & Altman, D. (1996). Whatever happened to the health insurance crisis in the United States? Voices from a National Survey. *Journal of the American Medical Association, 276*(16), 1346–1350.

Dutton, D. B. (1986). Social class, health and illness. In: L. Aikon & D. Mechanic (Eds), *Applications of Social Science to Clinical Medicine and Health Policy* (pp. 31–62). New Brunswick, NJ: Rutgers University Press.

Feldman, J. J., Makuc, D. M., Kleinman, J. C., & Cornoni-Huntle, J. (1989). National trends in educational differentials in mortality. *American Journal of Epidemiology, 129*(5), 919–933.

Friedman, E. (1997). Managed care, rating and quality: A tangled relationship. *Health Affairs, 6*, 29–47.

Furlong, B., & Wilken, M. (2001). Managed care: The changing environment for consumers and health care providers. *Research in the Sociology of the Health Care, 19*, 3–20.

Gallagher, T. J., Wagenfeld, M. O., Baro, F., & Haepers, K. (1994). Sense of coherence, coping, and caregiver role overload. *Social Science & Medicine, 39*(12), 1615–1622.

Ganzach, Y. (1998). Nonlinearity, multicollinearity and the probability of type II error in detecting interaction. *Journal of Management, 24*, 615–622.

Geyer, S. (1997). Some conceptual considerations on the sense of coherence. *Social Science & Medicine, 44*(12), 1771–1779.

Guralnik, J. M., Land, K. C., Blazer, D., Fillenhaum, G. G., & Branch, L. G. (1993). Educational status and active life expectancy among older blacks and whites. *New England Journal of Medicine, 329*(2), 110–116.

Hafferty, F. W., & Light, D. W. (1995). Professional dynamics and the changing nature of medical work. *Journal of Health and Social Behavior* (Extra Issue), 132–153.

House, J. S., Lepkowski, J. M., Kinney, A. M., Mero, R. P., Kessler, R. C., & Herzog, A. R. (1994). The social stratification of aging and health. *Journal of Health and Social Behavior, 35*(3), 213–234.

Jones, K. C. (1997). Consumer satisfaction: A key to financial success in the managed care environment. *Journal of Health Care Finance, 23*(4), 21–32.

Kivimäki, M., Feldt, T., Vahtera, J., & Nurmi, J.-E. (2000). Sense of coherence and health: Evidence from two cross-lagged longitudinal samples. *Social Science & Medicine, 50*(4), 583–597.

Kobasa, S. C. (1982). The hardy personality: Toward a social psychology of stress and health. In: G. S. Sanders & J. Suls (Eds), *Social Psychology of Health and Illness* (pp. 3–32). Hillsdale, NJ: L. Erlbaum.

Kronenfeld, J. J. (2001). New trends in the doctor-patient relationship: Impacts of managed care on the growth of a consumer protections model. *Sociological Spectrum, 21*(3), 293–317.

Lin, N. (2001). Building a network theory of social capital. In: N. Lin, K. Cook & R. S. Burt (Eds), *Social Capital: Theory and Research* (pp. 3–29). Hawthorne, NY: Aldine De Gruyter.

Link, B. G., & Phelan, J. (1995). Social conditions as fundamental causes of diseases. *Journal of Health and Social Behavior* (Extra Issue), 80–94.

Link, B. G., & Phelan, J. (2000). Evaluating the fundamental causal explanations for social disparities in health. In: C. E. Bird, P. Conrad & A. Fremont (Eds), *Handbook of Medical Sociology* (5th ed., pp. 33–47). Upper Saddle River, NJ: Prentice-Hall.
Lorber, J. (1997). *Gender and the social construction of illness*. Thousand Oaks, CA: Sage.
Margalit, M., Leyser, Y., & Avrabam, Y. (1989). Classification and validation of family climate subtypes in kibbutz fathers of disabled and non-disabled children. *Journal of Abnormal Child Psychology, 17*(1), 91–107.
Marmot, M. (1996). The social patterns of health and disease. In: D. Blane, E. Brunner & R. Wilkinson (Eds), *Health and Social Organization* (pp. 42–70). London: Routledge.
Marshall, N. L. (2001). Health and illness issues facing an aging workforce in the new millenium. *Sociological Spectrum, 21*(3), 431–439.
Mirowsky, J., & Hu, P. N. (1996). Physical impairment and the diminishing effects of income. *Social Forces, 74*(3), 1073–1096.
Mirowsky, J., & Ross, C. E. (1998). Education, personal control, lifestyle, and health: A human capital hypothesis. *Research on Aging, 20*(4), 415–449.
Mirowsky, J., Ross, C. E., & Reynolds, J. (2000). Links between social status and health status. In: C. Bird, P. Conrad & A. Fremont (Eds), *Handbook of Medical Sociology* (5th ed., pp. 47–67). Upper Saddle River, NJ: Prentice-Hall.
Pappas, G., Queen, S., Hadden, W., & Fisher, G. (1993). The increasing disparity in mortality between socioeconomic groups in the United States, 1960 and 1986. *New England Journal of Medicine, 329*(2), 103–109.
Pearlin, L. I., Leiberman, M. A., Menaghan, E. G., & Mullan, J. T. (1981). The stress process. *Journal of Health and Social Behavior, 22*(4), 337–356.
Pescosolido, B. A. (1992). Beyond rational choice: The social dynamics of how people seek help. *American Journal of Sociology, 97*(4), 1096–1138.
Pincus, T., Esther, R., DeWalt, D. A., & Callahan, L. F. (1998). Social conditions and self-management are more powerful determinates of health than access to care. *Annals of Internal Medicine, 129*(5), 406–411.
Pugliesi, K. (1995). Work and well being: Gender differences in the psychological consequences of employment. *Journal of Health and Social Behavior, 36*(1), 57–71.
Reynolds, J. R., & Ross, C. E. (1998). Social stratification and health: Education's benefit beyond economic status and social origins. *Social Problems, 45*(2), 221–247.
Robert, S. A., & House, J. S. (2000). Socioeconomic inequalities in health: An enduring sociological problem. In: C. Bird, P. Conrad & A. Fremont (Eds), *Handbook of Medical Sociology* (5th ed., pp. 79–97). Upper Saddle River, NJ: Prentice-Hall.
Robinson, J. C. (1997). Physician-hospital integration and the economic theory of the firm. *Medical Care Research and Review, 54*(1), 3–24.
Ross, C. E., & Bird, C. E. (1994). Sex stratification and health lifestyle: Consequences for men's and women's perceived health. *Journal of Health and Social Behavior, 35*(2), 161–178.
Ross, C. E., & Mirowsky, J. (1989). Explaining the social patterns of depression: Control and problem-solving – or support and talking. *Journal of Health and Social Behavior, 30*(2), 206–219.
Ross, C. E., & Mirowsky, J. (1992). Households, employment, and the sense of control. *Social Psychology Quarterly, 55*(3), 217–235.
Ross, C. E., & Mirowsky, J. (2000). Does medical insurance contribute to socioeconomic differentials in health? *Milbank Quarterly, 78*(2), 291–321.
Ross, C. E., & Wu, C.-L. (1995). The links between education and health. *American Sociological Review, 60*(5), 719–745.

Ross, C. E., & Wu, C.-L. (1996). Education, age, and the cumulative advantage in health. *Journal of Health and Social Behavior*, *37*(1), 104–120.

Rotter, J. B. (1966). Generalized expectancies for internal vs. extended control of reinforcements. *Psychological Monographs*, *80*, 1–28.

Seccombe, K., & Amey, C. (1995). Playing by the rules and losing: Health insurance and the working poor. *Journal of Health and Social Behavior*, *36*(2), 168–181.

Sofaer, S. (1998). Aging and primary care: An overview of organizational and behavioral issues in the delivery of healthcare services to older Americans. *Health Services Research*, *33*, 298–321.

Unland, J. J. (1998). The range of provider/insurer configurations. *Journal of Health Care Finance*, *24*(2), 1–35.

Vladeck, B. C. (1999). Managed care's fifteen minutes of fame. *Journal of Health Politics Policy and Law*, *24*(5), 1207–1211.

Ware, J. E., Jr., & Sherbourne, C. D. (1992). The MOS 36-item short form health survey (SF-36). I. Conceptual framework and item selection. *Medical Care*, *30*(6), 473–483.

Wellman, B., & Frank, K. A. (2001). Network capital in a multi-level world: Getting support from personal communities. In: N. Lin, K. Cook & R. S. Burt (Eds), *Social Capital: Theory and Research* (pp. 233–273). New York: Aldine De Gruyter.

Wilkinson, R. G. (1996). *Unhealthy societies: The afflictions of inequality*. London: Routledge.

Winkleby, M. A., Jatulis, D. E., Frank, E., & Fortmann, S. P. (1992). Socioeconomic status and health: How education, income, and occupation contribute to risk factors for cardiovascular disease. *American Journal of Public Health*, *82*(6), 816–820.

Wylie-Rosett, J., Walker, E. A., Shamoon, H., Engel, S., Basch, C., & Zybert, P. (1995). Assessment of documented foot examinations for patients with diabetes in inner-city primary care clinics. *Archives of Family Medicine*, *4*(1), 46–50.

THE INFLUENCE OF CLINIC ORGANIZATIONAL FEATURES ON PROVIDERS' ASSESSMENTS OF PATIENT ADHERENCE TO TREATMENT REGIMENS

Karen Lutfey

ABSTRACT

This study uses ethnographic data from two diabetes clinics to examine how some organizational features of medical settings are connected to the daily cognitive and interactional work of medical providers – specifically, the process of assessing patient adherence and using such assessments to make treatment decisions. I address continuity of care, scheduling and time constraints, team management, provider interaction, and medical recordkeeping as organizational-level issues that impact individual-level providers' work. More than a top-down model of how "macro" influences "micro," this study highlights how organizational influences are accounted for in terms of variation in patients' behavior.

Reorganizing Health Care Delivery Systems: Problems of Managed Care and Other Models of Health Care Delivery
Research in the Sociology of Health Care, Volume 21, 63–83

ISSN: 0275-4959/doi:10.1016/S0275-4959(03)21004-8

INTRODUCTION

In this study, I examine how some organizational features of medical settings – often considered the domain of "macrosociology" – are connected to the daily cognitive and interactional work of medical providers – often defined as "microsociology." As a linchpin for this investigation, I begin with the following observation, which is a robust theme in the data I bring to bear below: in the course of their work, clinical providers must be primarily concerned with accountable medical decision making, particularly in terms of designing treatment regimens that are effective but safe. To do this in the context of diabetes care, providers must make adherence assessments, or predictions of how closely their recommendations will be followed when patients leave the clinical setting. Clearly, practitioners' individual cognitive and philosophical styles are one facet of this assessment process, and this is a topic I elaborate elsewhere (Lutfey, 2003). Here, I use ethnographic and interview data from two diabetes clinics to analyze organizational features of health care delivery systems as another critical, yet poorly understood, influence on adherence assessment.

Organizational features of health care delivery systems affect, for example, which practitioners see which patients; how practitioners provide and acquire information about patients; how they record these assessments for future use by themselves or other practitioners who will be treating patients; and, ultimately, how they make and account for treatment decisions. Specifically, I consider: (1) How organizational features of clinical settings are integral to adherence assessments, and, by extension, the concept of patient adherence more generally; and (2) How the systematic influence of medical organizations is obscured when providers and researchers account for medical decision making in terms of individual patient behaviors such as adherence. In other words, I look to organizational features of medical settings for sources of variation in adherence assessment that are independent of variation in patient behavior, thereby contributing to a broader, multi-level understanding of health care delivery systems and patient adherence.

My analysis is focused on two Midwestern diabetes clinics in which the organizational characteristics influencing adherence assessment are, to varying degrees, salient to health care delivery systems throughout the United States. Through this type of analysis, we can examine everyday processes of medical care that are inaccessible via aggregated data and generally beyond the scope of studies focusing on organizations as units of analysis. Implications for health care delivery systems and sociological research on health care generally are numerous, and are discussed below.

BACKGROUND

Patient adherence, or the execution of medical recommendations,[1] is considered a linchpin in producing desired health outcomes. Accordingly, voluminous medical and social scientific literatures have mounted extensive research programs offering competing explanations for a narrow range of questions related to patients' behavior and why they do not follow medical advice. For example, when social scientific research into non-adherence first proliferated in the 1970s, it focused on individual patient characteristics, trying to discover how many people did not adhere to their regimens, who they were, and why they did not follow directions (Stimson, 1974, pp. 97–98). Various accounts for this behavior point to individual personality characteristics, which were seen as producing deviant behavior in patients: disliking side-effects of the drugs; having uncooperative personalities; being unable to understand physicians' instructions; and having a lack of motivation (Conrad, 1987, p. 15; Rosenstock, 1974; Stimson, 1974, p. 99; Svarstad, 1986, p. 440). More recent developments in sociological research suggest that there are generally logical explanations for patients' decisions not to follow instructions, such as difficulties in navigating the medical system (Becker et al., 1993; Hill, 1995); socioeconomic limitations (Dutton, 1986; Hill, 1995; Link & Phelan, 1995); or constraints arising from relationships with family members (Davis, 1991; Peyrot et al., 1987; Rajaram, 1997). While this research demonstrates rational aspects of non-adherence and portrays patients as competent, active agents instead of deficient, passive recipients of information, it still assumes that non-adherence is an objectively deviant, if understandable, quality of individuals. As a result, these agendas tend to systematically occlude entire domains of inquiry that are of interest to sociology and warrant further examination – such as the processes by which providers assess adherence in the first place.

Medical sociology also has a long history of using organizational theory to examine the complex organizations that are central to the delivery of health care. Most generally, organizational theory is concerned with how internal organizational structures function to motivate people and generate outcomes consistent with the goals of the organization; how external events impact internal workings of organizations; and how internal and external events can impact the survival of an organization (Fligstein, forthcoming). Applied to medical sociology, organizational theory has provided a backdrop for a wide range of topics including medical professions (Hafferty & Light, 1995; Light, 2000; Starr, 1982), medical education (Bloom, 1988; Bosk, 1979; Hafferty, 1991), managed care (Mechanic, 1998; Wholey & Burns, 2000), and the role of technology in medicine (Timmermans, 2000). In this context, individuals are of interest to the extent that they contribute to internal workings and reproduction of organizations, while less is known about the interfaces among organizations, individual providers, and patients. I explore the connection between organizational features of health care settings and the activities of providers as they contribute, not to institutional outcomes, but to decision making about and provision of services to individual patients.

This approach resonates with some earlier sociological work in non-medical settings. For example, in his 1974 work on "telling the convict code," Wieder (1974) describes how residents and staff of a halfway house in Los Angeles analyze and organize residents' behavior with reference to what he terms the "convict code." He suggests that is not that residents' behavior was driven or prescribed by the rules contained in the convict code, but that the intelligibility of any specific act could be seen as deriving from the code. In Wieder's words, " 'telling the code' rendered residents' behavior rational for staff by placing the acts in question in the context of a loose collection of maxims which compelled their occurrence" (1974, p. 156). In a similar way, practitioners rely on the notions of "rational decision making" and "patient adherence" as a rubric for interpreting phenomena in their work settings that other researchers, situated outside that immediate context of patient care, might understand in terms of organizational issues. Furthermore, practitioners use this orientation frequently, collaboratively, and in ways that enable them to justify or render accountable the work they have done. In short, practitioners are acutely interested in macrosociological aspects of medical settings, but they translate that information as they use it into something that is much more attuned to the specifics of their daily work.

This topic also resonates with Bittner's (1965) work on "the concept of organization." Bittner argues that organizational characteristics of bureaucracies exist not naturally "out there" in the world, but only insofar as actors bring that schema to bear on indexical phenomena. More importantly, however, Bittner argues that the concept of organization is a special schema in that once problems are referred to it, they acquire unique meanings and the organization has some power in invoking solutions to those problems. Applied to the current study, this means that the work of designing medical treatment regimens is fundamentally shaped by formal organizational settings and roles such as doctors, patients, clinics, and medical licenses. A doctor trying to design a diabetes treatment regimen is faced with a very different problem than a mother trying to figure out how best to take care of her child who has a cold – precisely because the former is seen as bound to a formal organization and its accompanying meanings and expectations for appropriate problem-solving activities. The availability of the "rule" that "safe but effective regimens must account for perceived patient adherence" derives from the notion of medicine as a formal organization – because doctors are legally liable for malpractice, they have licenses to maintain, and so on. In this sense, organizational features of health care settings become something intimately connected to providers engaged in the work of adherence assessment and accountable medical decision making.

DIABETES

Diabetes provides an excellent case for studying patient adherence and organizational features of health care settings. As a long-term illness whose treatment depends heavily on patient self-management, the study of diabetes may offer insights about the processes involved in patients and providers negotiating treatment regimens in the context of health care settings.[2] In diabetes care, patients are expected to independently manage complex daily treatment regimens involving medication, diet, and exercise in order to avoid acute and long-term difficulties. Chronically high glucose levels are correlated with serious complications that include circulatory precursors to amputation, blindness, kidney failure, heart disease, and stroke. Emulating normal glucose levels necessitates matching injections of insulin with the size, content, and timing of meals; ideally, people with diabetes should be eating low-fat, low-sugar, nutritious meals at regular intervals throughout the day. In addition, patients are expected to monitor their glucose levels by sticking their finger with a lancet and placing a drop of blood into a small electronic meter, and then record these figures in a log. Finally, they must monitor exercise, stress levels, and infections, each of which can cause fluctuations in glucose levels.

At its most simple, a diabetes regimen may consist of one injection of long-acting insulin, accompanied by limited diet restrictions. Regimens become more complex with the addition of multiple injections; mixing of long- and short-acting insulins; more extensive monitoring and assessing of food content; adjusting dosages of insulin according to algorithms; and increased glucose monitoring. At its most sophisticated, diabetes regimens involve insulin pumps, which most closely mimic healthy pancreatic activity. In Park and County[3] clinics, the settings under investigation here, patients usually began with fairly simple regimens and moved to more (or less) complex regimens, depending on practitioners' assessments of their success in managing glucose levels.

The profundity of the relationship between patient adherence and diabetes treatment extends beyond simplistic problems with non-adherence, however, and this is how organizational features and adherence assessment are critical. To make diagnostic adjustments to patients' insulin dosages that will continue to lower glucose levels without putting them at risk for acute (and potentially life-threatening) episodes of hypoglycemia, physicians must assess on each visit how closely patients are following their regimens. These assessments have very real health implications: patients who demonstrate tight control over their blood sugars are stronger candidates for insulin adjustments that continue to improve their glucose control and lower the likelihood of suffering long-term complications, while patients who do not have tight control are more likely to have regimens that minimize problems with hypoglycemia and result in higher overall glucose levels. Given that the majority of patients are unable to regularly follow all aspects of their regimens, practitioners are *continually* and *unavoidably* confronted with the task of assessing adherence in order to provide basic medical treatment for diabetes patients. Below, I elaborate several ways in which clinic organizational features shape this process.

DATA AND METHODS

The data for this study are from a year-long ethnographic study I conducted in 1997–1998. The fieldwork sites were two weekly subspecialty endocrinology clinics at two hospitals that are both part of the same University-based medical center located in a large, Midwestern city.[4] As such, the organizational features I analyze below are commensurate with many health care settings across the United States, particularly teaching hospitals, subspecialty clinics, and settings in which chronic illnesses are managed. The clinics were selected to provide an optimal contrast of the socioeconomic diversity of persons with diabetes: survey data collected from patients and providers, displayed in Table 1, reveal that those at County clinic are less likely to be white, are more likely to be uninsured, have lower incomes and education, and are rated by themselves and by providers as having poorer health and poorer diabetes control than patients at Park clinic.

The organization of personnel also differs between the two clinics. Park clinic has two endocrinologists and two nurse practitioners; there is also sometimes a single resident participating in the clinic as part of a four-week rotation. An adjacent Diabetes Education Center includes two full-time and one part-time certified diabetes educators, two dieticians, a medical social worker, and a full-time secretary. By contrast, County clinic is usually supervised by four attending physicians, but residents play a much larger role than in Park. All of the attending physicians in County are fully-credentialed endocrinologists with extensive experience in their fields, but, compared to the physicians at Park, they are more diverse in their specializations, spend a smaller percentage of their time seeing patients, and do not maintain their own ongoing patient caseloads. There are also two endocrinology fellows (advanced residents) who are appointed to County clinic for two years, and there are typically 2–4 residents attending the clinic in any

Table 1. Comparison of Patients Between Park and County Clinics.

	Park Clinic	County Clinic	*p*-Value for Difference
% Black/Hispanic	12	45	<0.001
Mean family income	$56,000	$12,000	<0.001
% Family income $15,000 or less	12	75	<0.001
% Without health insurance	3	42	<0.001
% College graduates	41	9	<0.001
% Less than high school education	11	36	<0.001
Patients' self-assessments of health (0–10 scale), 10 is most healthy	6.79	5.59	0.0015
Physicians' assessments of diabetes control (0–10 scale), 10 is best controlled	6.63	4.91	<0.001
N	137	33	

given week as part of a four-week rotation. Finally, there is also a certified diabetes educator who attends this clinic weekly on a volunteer basis. Compared to County, these personnel differences create an environment in Park with significantly higher continuity of care as well as a much more extensive and specialized center for diabetes education.

I collected several different types of ethnographic data. First, I observed approximately 250 hours of activity at these clinics, including approximately 200 different consultations between diabetes patients and medical practitioners. Second, I videotaped over twenty hours of these consultations. Third, I conducted semi-structured qualitative interviews with 25 practitioners, including all of the physicians in the University medical center who treat diabetes, as well as nurses, dietitians, social workers, and diabetes educators. Fourth, I conducted focus-group interviews with three diabetes support groups. Paid assistants transcribed these audio- and video-taped data, after which I conducted detailed analysis using qualitative data analysis software. Fifth, I conducted brief telephone surveys with 170 diabetes patients in order to collect basic demographic information, as well as information on patients' beliefs about diabetes, their reported reasons for not always following the instructions given by their doctors, and their expenses related to the disease. This telephone sample comprises 86% of all of the diabetes patients seen at the two clinics over a three-month period, which was the typical interval between clinic visits for diabetes patients. Sixth, for each of the patients surveyed, attending physicians (and, when applicable, medical residents) completed short questionnaires about patient adherence.

AVAILABILITY OF PERSONNEL: WHICH PROVIDERS SEE WHICH PATIENTS?

Concerns about continuity of care in health care are widespread in the United States, both explicitly in debates over managed care versus fee-for-service and implicitly in research on trust in physicians. Classic work in the area of medical sociology addresses continuity of care in the context of physician authority and patient utilization of medical services(Freidson, 1988; Parsons, 1951; Waitzkin, 1991). For the practitioners in my study, however, concerns about continuity of care were acted on and discussed in the context of patient adherence. They were challenged to design treatment regimens that would be effective yet safe for patients – but if they were not able to see the same patient repeatedly, it became difficult to accurately assess their situations, as one physician explained in an interview:

> Another physician might say, "Well, you know, that patient's not very compliant." And that's their view. Maybe you have a different view. Or maybe they just haven't been given the proper education. Or maybe they've been bounced around so much and they keep hearing different things and they don't know what to think, so they're getting frustrated. The more you have continuity or follow-up with that patient, the more you're going to get a more accurate picture of what they do and don't understand and how compliant they may or may not be.

Organizationally speaking, this physician is talking about continuity of care. This feature of the clinic, however, is most immediately intelligible to him in terms of his medical decision making and adherence assessment of patients. Part of his work of doing responsible, rational, accountable medicine when he is in clinic, is getting an "accurate picture" of how adherent patients might be so that he can design an appropriate treatment. In this context, patients bouncing around to different practitioners is most properly understood vis-à-vis adherence and regimen design.

Juxtaposed against Park, County's heavy reliance on residents, students, fellows, and rotating attending physicians decreases continuity of care, and highlights how the organization of personnel available to see patients impacts the process of adherence assessment. To the degree that adherence assessments are treated as neutral conduits for understanding and predicting patients' "true" behavior, the influence of organizational features on individual decision-making is obscured by the rhetoric of patient adherence. During one interview, which occurred just after a clinic meeting, an attending physician from County discussed a patient who had been seen that morning. This patient's lab results indicated a low average blood sugar level, but the practitioners suspected that this value resulted from many highs and lows as opposed to good overall glucose control. When

I asked about the assumptions practitioners in the clinic make about treatment adherence based on lab results, he responded:

> Ideally, (t)he outcome is not based on a single test. Nor a single blood sugar, nor a single week, or a single visit, or a single anything . . . The outcome is based on the totality of the things that you assess over the period of time that you see the patient. (By contrast,) (t)he focus that you see in the clinic like we have is that the totality of the assessment comes down to "What's your last Hemoglobin A1C?" Which doesn't mean jack.

As a result, providers who do not see patients regularly are forced to rely more heavily on laboratory indicators such as the Hemoglobin A1C, a blood test providing average glucose levels for the previous three months. While providers are certainly cognizant of low continuity of care as an organizational problem, this feature is most relevant to their work as it inhibits their collection of information pertaining to treatment decisions.

The relatively low continuity of care at County also impacts adherence assessments *over time* because so many of the practitioners cannot be sure that they will have the opportunity to closely monitor or follow patients they treat, and do not necessarily know which practitioner will see the patient on his next clinic visit. Therefore, they are constrained in their ability to try new treatments or to pursue regimens that might increase the likelihood of negative side effects such as hypoglycemia.[5] Yet, when providers talked about these decision-making processes – which residents did explicitly in the course of presenting cases to attending physicians – they never accounted for these decisions in terms of organizational features of the health care delivery system in which they were working. While providers clearly understand such organizational constraints, that orientation is not treated as a useful rubric for accounting for work practices. "Low continuity of care" may indeed be a critical underlying reason for a resident deciding to prescribe a conservative treatment regimen, but the more immediate reason, "having insufficient information to predict whether a more aggressive regimen will be safe based on a patient's adherence behavior," is the one that is considered accountable in the immediate clinical context.

Responding to this feature of clinics in terms of medical decision-making and adherence assessment is not just a matter of personal preference or a way of describing one physician's decision-making – providers often spoke about continuity of care as an organizational problem in interviews, but oriented to it as a limitation in adherence assessment in the in situ reality of treating patients. It is in this sense that researchers have limited understanding of continuity of care as part of a rubric for patient assessment and decision-making. Indeed, continuity of care serves as a conceptual umbrella for several of the topics addressed below (see also Lutfey & Freese, 2003).

STRATEGIES FOR CARE: HOW DOES THE FORM OF TREATMENT AFFECT ITS CONTENT?

Scheduling and Time Constraints

Increased standardization of scheduling and time constraints has also been a topic of interest for organizational sociologists, particularly in the context of managed care, third party payers, and the historical emergence of clinics like the one I studied (Mechanic, McAlpine & Rosenthal, 2001). These constraints are addressed as predictors of patient satisfaction or health outcomes, but less is known about their roles in providers' accounts for this work of assessing adherence. These time constraints alone are potentially troublesome for practitioners' work, as one physician explained to me:

> The patients will tell you, "My doctor's too busy to listen to me or talk to me. I mean, they're so busy they only have ten minutes," or something. I mean, my God, what can you say in ten minutes?

Furthermore, practitioners' abilities to maintain their patient schedules during a given clinic are subject to a wide variety of variables which may or may not be under their control: whether patients are able to arrive on time for their appointments; unforeseen medical, social, psychological, or interactional circumstances requiring practitioners to spend more time than was scheduled with a given patient; slowdowns related to inexperienced residents seeing patients; delayed availability of attending physicians to consult with residents about cases; or delays created by practitioners needing to make phone calls or look up patients' records to make treatment decisions.

In Park clinic, for example, practitioners are generally scheduled to see one patient every thirty minutes. When their schedules are delayed by a few minutes with each patient, the cumulative effects are noticeable, particularly given busy afternoon schedules and inability to extend morning clinic appointments past the lunch hour. Pressure to maintain appointment schedules can be related to practitioners' assessments of patient adherence insofar as time constraints are often antithetical to the time-intensive process of collecting information from patients to more accurately assess their behavior (a problem exacerbated by low continuity of care). One practitioner, for example, discussed her frustration in dealing with patients she views as unwilling to entertain medical advice:

> It's hard to keep from feeling like, "Oh, you're wasting my time." It's like, if you really and truly are clearly not going to even think about some of the things that we've talked about, what are you doing here? I have a whole line of other people waiting to see (me) that may want to implement some of the recommendations.

In a system that treats practitioners' time as an important commodity, however, tight scheduling can be counteracted by efficiency in interaction. In other words, patients who are responsive to medical recommendations, and who display this responsiveness in time-saving ways (e.g. by having memorized their medication dosages) are more readily assessed as adherent than those patients who "waste" practitioners' time (e.g. those who cannot remember their medications, thereby making it difficult to confirm regimens listed in a patient's chart on a previous visit). In these instances, organizational-level time constraints imposed by the clinic are rendered intelligible by individual providers in terms of information that can be obtained about a patient and how that information can be used to make a rational medical decision.

Team Management

Diabetes care often involves a team of specialized practitioners to address the multiple physical aspects of diabetes, the need for extensive patient education, and the need for frequent communication with patients. In fact, the 1997 logo for the American Association of Diabetes Educators proclaimed, "Self-Management Matters: Team up with a Diabetes Educator," promoting increased patient efficiency as an outcome of team approaches. One practitioner characterized the movement toward self-management as follows, framing it as a broad organization-level change across health delivery systems:

> The attitude of having patients take care of themselves is not only in diabetes, but it's in all kinds of things right now to save money for the HMO and the PPO. I think it is just a new approach trying to get the patient to do more and the medical providers to do less and I think that it's maybe even economically motivated.

The importance of a team approach in diabetes care was also a regular theme in practitioner interviews, highlighting additional connections between organizational characteristics and care for individual patients (Light, 1988). The following provider, in fact, explicitly downplayed the role of individual patients as a major problem in comparison to organizational-level difficulties:

> The biggest challenge in taking care of diabetics in the systems that I've had to deal with, even in the clinic we have now, is not the disease and not the patient, but rather the system . . . So the biggest problems we have (in County) and in most places is (that) it's a disease that requires multiple, technical approaches – dietitians, people interested in feet, people who want to teach people how to take insulin, and somebody who juggles these things.

Similarly, several other practitioners claimed during interviews that the biggest challenges they face in treating diabetes patients are the problems in coordinating this team approach: finding sufficient time and personnel resources for patients; communicating among members of the team; and documenting patient care in such a complex organization, particularly given the high rates of practitioner turnover in teaching facilities.

In the context of patient care, however, these organizational factors are implicated as part of the logic and accountability involved in assessment of and decision making about individual patients. That is, problems at the level of everyday practices in the clinics are not accounted for in terms of economic motivations of HMOs, or as unfortunate outcomes of insufficient funds in publicly-funded clinics, but as an absence of strong adherence displays from patients, which constrains safe and accountable treatment options. In the same way that Wieder's convicts accounted for their behavior as if it was prescribed by rules in the convict code, providers in Park and County accounted for their medical decision making in terms of patient adherence, and which regimens were safe-yet-effective for particular patients. Even when assessments of adherence were impeded by underdeveloped "teams," as in County hospital, those problems were never discussed *in situ* as organizational problems.

For several reasons, patients involved in team management were more likely to have well-controlled diabetes. Specialists (e.g. podiatrists, nephrologists) could monitor their complications, provide regular dietary education, and offer other similar types of services. Patients who were treated by a team often developed skills for managing their diabetes and the medical system that other patients did not possess. As they continued to participate in team management and improve these skills, their effectiveness at managing diabetes – and conveying to practitioners that they knew what they were doing – increased. Furthermore, based on plain economic differences, patients had differential access to team management resources; patients using the County clinic, for example, did not have access to the Diabetes Education Center in Park, which offered more centralized and specialized diabetes education. Low continuity of care exacerbates differential access to team management: in cases where doctor-patient relationships are not well-established, it is difficult to provide services that are tailored to patients' specific needs. Differential access to resources, then, is more than a simple matter of what individual patients can afford, and instead, access is also connected to organizational features of clinics that are independent of individual patient characteristics. The distinction between patient characteristics and constraints imposed by organizational features, however, was not articulated when practitioners account for their treatment decisions. As a result, organizational features resulting in differential access to medical resources were often oriented to by researchers in terms of variation in providers' treatment behaviors or patients' adherence.

ORGANIZATIONAL CULTURES: WHAT HAPPENS WHEN PROVIDERS INTERACT WITH ONE ANOTHER?

Clinical Interactions Among Practitioners

Provider assessments of patients are based on extensive, cumulative social interactions, even for new patient-practitioner relationships. Therefore, the question of how information is retained, organized, transmitted, and utilized is important for all health care providers, not only those involved in teaching or treating chronic illnesses such as diabetes. In County and Park clinics, for example, practitioners share common physical space that facilitates their interaction when they are not seeing patients. It was not at all uncommon for practitioners in both clinics to discuss specific patients, often at length; these interactions were also extremely varied, ranging from discussion of remarkable events in patients' lives to their adherence with regimens and prognoses for serious complications. While there was significant variation in the actual terms that were used (e.g. "compliance," "adherence," or, occasionally, "personality disorder"), attributions about patient behavior were extremely common. I argue that these interactions constitute an important arena for shaping practitioners' assessments of patient adherence both in terms of developing informal ideas about the generic role of adherence in medical care and by collectively assessing specific patients.

Furthermore, informal interactions often contributed to assessments of specific patients, since patients in Park and County were seen by multiple practitioners on a given visit (by residents or nurse practitioners prior to an attending physician, for example), or by multiple practitioners over the course of their history at the clinic. Adherence assessments made by single practitioners did not occur in a social vacuum: in the same way these informal discussions conveyed relevant information about formal regimen design and medical treatment, they also conveyed information salient to adherence assessments. The following example illustrates how informal practitioner discussion of a patient's adherence seems related to subsequent interaction that patient had with two practitioners. The patient in this case was a white male named Bill, who was in his 50s with advanced complications from his diabetes. The nurse practitioner who normally treats him, Julie, brought up the possibility of Joe, a resident currently rotating through the clinic, seeing Bill first because he was an "interesting" patient with incurable complications. Julie told us in advance that the patient was not very "with it," despite having a wife who was contrastingly articulate and who did seem fairly "with it." She also implied that he did not do the things he was supposed to do to take care of himself and that he neglected his health, but phrased this in terms of it being "hard to get people to change their ways," in a sympathetic way. Her stated goal for Bill was to try to keep things from getting any worse, but it seemed clear that his diabetes complications would eventually be fatal.

In his consultation with Bill, which I observed, Joe spent most of the time reviewing the patient's complications and monitoring how they had changed. At the outset of the conversation, he asked Bill, "Do you stick with the diet pretty well?" and Bill nodded and responded that he did. Joe nodded back, and added, "So you are watching what you eat, eating at the right times?" Through their construction, both of these queries solicited positive responses from the patient, confirming that he was indeed adhering to his diet (Pomerantz, 1984). Throughout the interaction, Joe never asked any open-ended questions about how closely Bill followed the diet or other recommendations, and Bill agreed with all the positive statements about adhering to the regimen. On Bill's previous clinic visit, however, his Hemoglobin A1C result was 12, and on this visit it was 9, which, while lower, is still beyond the normal range (which is 4–6). Despite this evidence of high average blood sugars, the presence of complications corroborating the notion of poor glucose management, and Julie's assertion that Bill did not follow his regimen closely, Joe solicited only confirmation that the patient was following his regimen, and accepted these responses without further inquiry or challenge. There is not evidence here to support the notion that Julie's comments prior to this consultation caused Joe to solicit and accept claims of diet adherence despite extensive evidence to the contrary. However, the line of questioning pursued by the resident was highly consistent with her opinions that the patient was unwilling to change, he was not "with it," and his complications had progressed to a point where diet adherence was moot. In contrast, a patient seen in the same clinic during the previous week, also riddled with serious complications but characterized by the attending physician as a "lovely" woman, was questioned extensively by a resident about her diet when her claims about her own behavior were incommensurate with her lab results.

After seeing the patient, Joe left the exam room and returned to the physicians' room to present the case to Julie and another attending physician. They had great difficulty determining the best course of action because Bill did not know which medications he was taking. The practitioners made a series of jokes challenging one another to be the practitioner who could "cure him." Julie then returned to the examination room to request that Bill call back in the afternoon with a list of his medications. In the room, Julie also told Bill that she wanted him to bring back all of his medications in a bag each time he came to the clinic so they could see what he was taking. He responded, "I can do that," and Julie jokingly said, "I know you can." He then made a joke about how poorly he was taking care of himself, at which point Julie made eye contact with him and said (still jokingly), "Well, I didn't say it." Bill responded with a laugh and said, "Yeah, but you didn't have to agree with me." Here, in the course of one clinic visit, a single practitioner's assessment of a patient's adherence was shared in different ways with multiple practitioners and with the patient himself, and yet these data do not even address the ways these assessments are shared over the course of multiple clinic visits or how they are reflected in other ways, such as medical history notes in the patient's chart. Furthermore, these data do not speak to the potential impact of the provider conveying her assessment to the patient, and how that might influence his attitudes or behavior regarding his diabetes management or healthcare. Minimally, in this case, the practitioner's assessment of the patient as highly non-adherent affected the focus of the treatment he received and others' perceptions of his intellect and personality (including that of the resident, who had never met the patient before) – findings which undermine research models treating adherence assessment as a neutral conduit for understanding "objective" patient behavior.

Medical Charts and Recordkeeping

At least as central to diabetes care as cognitively- and interactionally-based assessments of patient adherence is the information stored in patients' charts and other medical records. The complexity of patient-practitioner interaction is distilled and recorded in these records on each clinic visit, and used, particularly in teaching settings, by a wide range of practitioners as tools for treating patients. In his discussion of " 'good' organizational reasons for 'bad' clinic records" in *Studies in Ethnomethodology*, Garfinkel (1967) analyzes connections between information recorded in clinic records and the organizational needs and demands of the clinical setting. His analysis highlights the "economic" tension between a need for information to be recorded and the issue of "How much of the nurse's (or the resident's or the social worker's, etc.) time will it take?" Furthermore, he argues, archiving is not a question of "Is this information worth the cost?" but "Will it have been worth the cost?" These principles apply to the process of record-keeping in Park and County clinics as well.

First, an analysis of decision making about which information to record reveals its fundamentally social character. While medical charts contain lab results and vital signs that are "objective" and largely shielded from social interpretation, there are also vast amounts of information which are identified, interpreted, and recorded in and through social interaction. The following excerpt from my field notes offers one example of how patient-practitioner interaction can translate to a medical record:

> The patient in this case is a white male professor, approximately 50-years old. The attending physician and I saw him alone. When the physician asked if he was following his diet, the patient said, "You know, yes and no," smiled, and turned to wink at me, laughing a little bit. When I saw his notes later, the physician had written "good compliance with diet" in the chart.

In this case, drawing the conclusion that the patient is following his diet closely depends minimally on the physician interpreting the patient's use of humor and ambiguity in his response to the initial question, and likely combines this information with other types of existing knowledge: their shared interpersonal history from previous clinic visits; previous and current lab results and the extent to which they corroborate his comment; previous documentation about the patient's behavior (the physician's own documentation or from other practitioners who have seen the patient); and his experience in treating other diabetes patients who communicate in different ways about their own adherence. This assortment of information provides a backdrop for the physician to understand the current interaction and make a treatment decision. To interpret medical records as abstracted from these social processes is to underappreciate the implications of clinic interactions for adherence assessments.

The "economic" character of decisions about recording information is particularly salient in the case of patient adherence. Ultimately, there is a low return for practitioners on recording extensive information about patient adherence, particularly with regard to the reasons underlying patients' behavior with their regimens. In a practical sense, the reasons for patient non-adherence are of less concern than assessing the actual degree to which they are following their regimens. While treatment decisions can be directly affected by the latter, speculation about the former is much more unstable and has fewer direct implications for treatment decision making. As one physician indicated in our interview when I asked about his biggest concerns in treating diabetes patients,

> Obviously the medical issues. I have to be concerned about them. I have an ethical and legal obligation to deal with those medical issues.

On a practical level, there are organizational reasons for practitioners to subscribe to traditional, parsimonious models of patient non-adherence rather than the complex, multi-level explanations that have been offered by social scientists in recent years. While many of the practitioners in these settings were quite knowledgeable about and sympathetic with socioeconomic constraints on patient behavior, the pursuit and recording of detailed, in-depth accounts for patient nonadherence is tangential to, and sometimes in opposition to, their more immediate professional medical goals.

While the majority of existing research on patient adherence has given little attention to the role of practitioners in assessing adherence, research that has addressed the influence of practitioners in defining medical situations has focused on patient-practitioner dyads or discussed these patterns as psychological phenomena without addressing the actual social mechanisms by which practitioners develop their ideas or a common culture (Fineman, 1991). My analysis attempts examine some of the broader connections between practitioners' individual approaches to adherence and the organizational features of the clinical settings in which they are embedded (Sudnow, 1967).

IMPLICATIONS FOR HEALTH CARE DELIVERY SYSTEMS

In this paper, I have argued that, even if it was possible to control for variation in patient behavior, organizational features of clinical settings would continue to impact providers' perceptions of patient adherence and their predictions of the extent to which patients will execute recommendations. While the current analysis is focused on data from two clinics, all the organizational characteristics discussed here are fundamentally common, to varying degrees, to health care delivery systems nationally. Specifically, I address continuity of care, scheduling and time constraints, team management of diabetes, provider interaction, and medical recordkeeping as organizational-level issues that influence individual-level providers' behavior. More than a top-down model of how "macro" influences "micro," however, this study highlights how *organizational* influences are accounted for, and rendered intelligible, in terms of variation in *patients' behavior*.

One way of thinking about this topic is to use Garfinkel's (1988) analogy comparing scientific study to the process of extracting an animal from foliage. Referring specifically to the processes by which astronomers identify a pulsar, Garfinkel argues that doing the work of identifying a pulsar from among all other available data, and to do that in a way that is demonstrably accountable to norms of proper scientific methodology, entails a broad range of artful and jointly accomplished activities on the part of the scientists. In a similar way, medical providers are faced with jungles of information in their work, and the animal they are trying to extract from that is legitimate, accountable medical decision-making. A basic mechanism for accomplishing this work is adherence assessment, which allows providers to rationally account for their treatment decisions as predictably safe and reasonably effective. It is not that the sorts of large-scale organizational features typically studied by "macro" sociology are irrelevant or disconnected from their work, or that providers themselves are unaware of organizational issues. It is more the case that, as part of their *in situ* work, providers orient to these features of their environments in ways that are systematically different from approaches taken by some researchers – and these orientations are different for reasons that are critical for medical science. Problems posed by organizational features of health care delivery systems simply do not offer accountable reasons for making particular treatment decisions, and therefore do not facilitate the everyday work of extracting animals from the foliage.

A major implication of this focus on patients is that medical practitioners and health services researchers alike may be more likely to attribute shortcomings or failures of medical treatment, including aspects of the health-SES gradient, to *individuals* instead of *systematic weaknesses in organizations*. Despite extended efforts focused on reorganizing health care delivery systems and trying to improve health outcomes, some of the processes identified in this paper may be perceived by organizational researchers as falling outside the scope of their interests, precisely because problems are ultimately framed as issues related to patient adherence. Alternatively, studies focused of the sorts of organizational features I discuss here, such as continuity of care and time constraints on providers, may not be equipped to fully appreciate the range of nuanced implications organizational features have for provider assessments, decision-making, and health outcomes. For effective study of health care delivery systems, we need to continually consider the connections between such "micro" and "macro" levels of analysis.

A second major implication derives from literature on patient adherence. Insofar as patient adherence is a central concern in health services, and existing literature has been largely unsuccessful in identifying the reasons underlying poor adherence, this study also implies that organizations may offer another venue to examine for answers. This is not to say, of course, that there is no objective variation in patients' health behaviors, or that we no longer need research focusing on patients, but that our knowledge about the connections among providers, patients, and health care delivery systems should be further developed. In the same way that organizational researchers may overlook effects labeled as "adherence issues," so may researchers interested in patient behavior purposely yet unnecessarily exclude study of organizational features of medical settings.

More generally, this study calls attention to our need, as sociologists, to focus attention on relationships between individuals and institutions, and the ways variation in institutional services can be rhetorically attributed to individual behavior. These findings also illustrates how theoretically fertile this area of investigation may be, and highlights the importance of exploring multi-level facets of health care delivery systems.

NOTES

1. The term "compliance" was first popularized in the 1970s, defined as "the extent to which individuals' behavior (in terms of taking medications, following diets, or executing lifestyle changes) does not coincide with medical or health advice" (Haynes et al., 1979). Subsequent research has advocated the term "adherence" as a replacement, arguing that it invokes a less paternalistic approach to healthcare (Lutfey & Wishner, 1999).

2. Diabetes is also sociologically relevant for several reasons. It is the seventh leading cause of death in the United States, and it disproportionately afflicts disadvantaged populations such as elderly people and minorities; more women suffer from diabetes than men, and more women die each year from diabetes than from breast cancer(Center for Disease Control and Prevention, 2002; Geiss, 1995). The costs associated with diabetes care are also extremely high, amounting to $44.1 billion in direct medical costs and an additional $54.1 billion in indirect costs (CDC, 2002). Diabetes is also increasing in prevalence, due partly to growing elderly and minority populations and to increasing rates of obesity.

3. Park and County are pseudonyms.

4. Because most diabetes patients in the United States are treated by practitioners of internal or family medicine as opposed to the specialty clinics studied here, my data include a disproportionately high number of people using insulin pumps and other sophisticated regimens compared to the general population. This bias works to my benefit in examining adherence assessments, however, since data collected from a general practitioner's clinic would have less variation in regimen types and provide less information how practitioners decide among them.

5. While neither hospital had collected official data on average patient glucose levels, discrepancies between the clinics, with Park having lower averages, were treated as general knowledge by most of the practitioners with whom I spoke. As a general indicator of these differences, I did not observe any patients with insulin pumps in County hospital and fewer than eight patients with Hemoglobin A1C results below 8 (where 4–6 is normal). By contrast, Hemoglobin A1C results in the double-digits were somewhat frequent in County and almost non-existent at Park.

ACKNOWLEDGMENTS

This research has been generously supported by funds from the Robert Wood Johnson Foundation, the University of Minnesota, and the American Association of University Women. I am grateful to Bernice Pescosolido, Doug Maynard, Bill Corsaro, Shel Stryker, and Anne Rawls for their input on earlier versions of this manuscript.

REFERENCES

Becker, G., Janson-Bjerklie, S., Benner, P., Slobin, K., & Ferketich, S. (1993). The dilemma of seeking urgent care: Asthma episodes and emergency service use. *Social Science and Medicine*, *37*, 305–313.

Bittner, E. (1965). The concept of organization. *Social Research*, *32*(3), 239–255.

Bloom, S. W. (1988). Structure and ideology in medical education: An analysis of resistance to change. *Journal of Health and Social Behavior*, *29*, 294–306.

Bosk, C. L. (1979). *Forgive and remember: Managing medical failure*. Chicago: University of Chicago Press.

Center for Disease Control and Prevention (2002). *National diabetes fact sheet: National estimates and general information on diabetes in the United States*. Atlanta: United States Department of Health and Human Services, Center for Disease Control.

Conrad, P. (1987). The non-compliant patient in search of autonomy. *Hastings Center Report August*, 15–17.

Davis, F. (1991). *Passage through crisis: Polio victims and their families*. New Brunswick: Transaction.

Dutton, D. (1986). Social class, health, and illness. In: L. Aiken & D. Mechanic (Eds), *Applications of Social Science to Clinical Medicine and Health Policy*. New Brunswick: Rutgers University Press.

Fineman, N. (1991). The social construction of non-compliance: A study of health care and social service providers in everyday practice. *Sociology of Health and Illness*, *13*, 354–374.

Fligstein, N. (forthcoming). Organizations: Theoretical debates and the scope of organizational theory. In: C. Calhoun, C. Rojek & B. Turner (Eds), *Handbook of Sociology*. New York: Russell Sage.

Freidson, E. (1988). *Profession of medicine*. Chicago: University of Chicago Press.

Garfinkel, H. (1967). *Studies in ethnomethodology*. Englewood Cliffs, NJ: Prentice-Hall.

Garfinkel, H. (1988). Evidence for locally produced, naturally accountable phenomena of order, logic, reason, meaning, method, etc., in and as of the essential quiddity of immortal ordinary society, (I of IV): An announcement of studies. *Sociological Theory*, *6*(1), 103–109.

Geiss, L. (1995). Are women more likely to die of diabetes than breast cancer? *Diabetes*, *44*(Suppl. 1), 123A.

Hafferty, F. (1991). *Into the valley: Death and the socialization of medical students*. New Haven: Yale University Press.

Hafferty, F. W., & Light, D. W. (1995). Professional dynamics and the changing nature of medical work. *Journal of Health and Social Behavior* (Extra Issue), 132–153.

Haynes, R. B., Taylor, D. W., & Sackett, D. L. (Eds) (1979). *Compliance in health care*. Baltimore: Johns Hopkins University.

Hill, S. (1995). Taking charge and making do: Childhood chronic illness in low-income black families. *Research in the Sociology of Health Care*, *12*, 141–156.

Light, D. W. (1988). Toward a new sociology of medical education. *Journal of Health and Social Behavior*, *29*(December), 307–322.

Light, D. W. (2000). The medical profession and organizational change: From professional dominance to countervailing power. In: C. E. Bird, P. Conrad & A. Fremont (Eds), *Handbook of Medical Sociology* (pp. 201–216). Englewood Cliffs, NJ: Prentice-Hall.

Link, B., & Phelan, J. (1995). Social conditions as fundamental causes of disease. *Journal of Health and Social Behavior* (Extra Issue), 80–94.

Lutfey, K. (2003). *Provider roles in maximizing patient adherence to medical treatment regimens*. Unpublished manuscript.

Lutfey, K., & Freese, J. (2003). *Toward some fundamentals of fundamental causality: Socioeconomic status and health in the routine clinic visit for diabetes*. Unpublished manuscript.

Lutfey, K., & Wishner, W. A. (1999). Beyond compliance is adherence: Improving the prospect of diabetes care. *Diabetes Care*, *22*(4), 635–639.

Mechanic, D. (1998). Emerging trends in mental health policy and practice. *Health Affairs*, *17*(6), 82–98.

Mechanic, D., McAlpine, D. D., & Rosenthal, M. (2001). Are patients' office visits with physicians getting shorter? *New England Journal of Medicine*, *344*(3), 198–204.

Parsons, T. (1951). *The social system*. Glencoe: Free Press.

Peyrot, M., James, F., McMurry, J., & Hedges, R. (1987). Living with diabetes: The role of personal and professional knowledge in symptom and regimen management. *Research in the Sociology of Health Care*, *6*, 107–146.

Pomerantz, A. (1984). Agreeing and disagreeing with assessments: Some features of preferred/dispreferred turn shapes. In: J. M. Atkinson & J. Heritage (Eds), *Structures of Social Action: Studies in Conversation Analysis* (pp. 57–101). New York: Cambridge University Press.

Rajaram, S. S. (1997). Experience of hypoglycemia among insulin dependent diabetics and its impact on the family. *Sociology of Health and Illness*, *19*(3), 281–296.

Rosenstock, I. (1974). Origins of the health belief model. *Health Education Monographs*, *2*, 378.

Starr, P. (1982). *The social transformation of American medicine*. United States of America: Basic Books.

Stimson, G. (1974). Obeying doctor's orders: A view from the other side. *Social Science and Medicine*, *8*, 97–104.

Sudnow, D. (1967). *Passing on: The social organization of dying*. Englewood Cliffs, NJ: Prentice-Hall.

Svarstad, B. L. (1986). Patient-practitioner relationships and compliance with prescribed medical regimens. In: L. Aiken & D. Mechanic (Eds), *Applications of Social Science to Clinical Medicine and Health Policy* (pp. 438–459). New Brunswick: Rutgers University.

Timmermans, S. (2000). Technology and medical practice. In: C. E. Bird, P. Conrad & A. M. Fremont (Eds), *Handbook of Medical Sociology* (pp. 309–321). New York: Prentice-Hall.

Waitzkin, H. (1991). *The politics of medical encounters*. New Haven: Yale University Press.

Wholey, D. R., & Burns, L. R. (2000). Tides of change: The evolution of managed care in the United States. In: C. E. Bird, P. Conrad & A. M. Fremont (Eds), *Handbook of Medical Sociology* (pp. 217–237). New York: Prentice-Hall.

Wieder, D. L. (1974). *Language and social reality: The case of telling the convict code*. The Hague: Morton.

PART II:
SPECIAL GROUPS OF PATIENTS AND HEALTH ISSUES

INNER STRENGTH AND THE EXISTENTIAL SELF: IMPROVING MANAGED CARE FOR HIV+ WOMEN THROUGH THE INTEGRATION OF NURSING AND SOCIOLOGICAL CONCEPTS

Joseph A. Kotarba, Brenda Haile, Peggy Landrum and Debra Trimble

ABSTRACT

The purpose of this paper is to contribute to the understanding of women's experiences of living with and surviving HIV/AIDS. We argue that strong conceptualization of this experience will lead to more efficient health care delivery for this growing population, within the general framework of managed care. Our analytical strategy is to integrate the nursing concept of inner strength with ideas from the sociological concept of the existential self. There are numerous definitions of the increasingly popular concept of inner strength in the health care literature, largely developed through the experiences of women living with breast cancer. In general, this concept is useful because it focuses research attention on patients' experiences and

Reorganizing Health Care Delivery Systems: Problems of Managed Care and Other Models of Health Care Delivery
Research in the Sociology of Health Care, Volume 21, 87–106

ISSN: 0275-4959/doi:10.1016/S0275-4959(03)21005-X

perceptions of illness. Nevertheless, current definitions can be critiqued for their tendency: (1) view inner strength as a thing-like phenomenon, as if it were like a disease, to be measured, treated and supplemented; (2) describe inner strength in overly metaphoric and romanticized terms that do not reflect the everyday life of living with a serious illness; and (3) assume that inner strength is equivalent to doing well. We argue that this concept can be of greater scholarly and clinical use if it is defined as follows: Inner strength refers to the different ways women with serious illnesses experience and, subsequently, talk about the deepest, existential resources available to and used by them to manage severe threats to body and self. We developed this concept through a series of 19 biographical and conversational interviews with women living with HIV/AIDS. Our interviews found that these women describe their experiences in terms of three types of narratives or stories. Faith stories recount the ways reliance upon a higher power (spiritual or religious) provides a sense of inner strength. Character stories recount the ways women experience inner strength as a resource available to them before as well as during their illness. Uncertainty stories recount the ways women perceive their inner strength as problematic. We conclude with specific suggestions for the application of our revised concept of Inner strength to the role of nursing in the delivery of managed care to women living with HIV/AIDS.

INTRODUCTION

Women comprise the fastest growing population of new cases of human immunodeficiency virus (HIV) infection worldwide (Wofsy, 1995). By the end of 2001, a total of 141,048 female adolescents and adult women in the United States had been reported to have acquired immunodeficiency disease (AIDS) or HIV (HIV/AIDS Surveillance Report, December, 2002). Between 1990 and 1994, the rate of increase in incidence of AIDS in women was 89% compared to a 29% increase reported in men. Clearly, the HIV/AIDS epidemic in women in the United States is escalating. Women of color are particularly at-risk for HIV infection, as are women who are partners of injecting drug users, younger women and women outside major population centers such as those in rural Southeastern United States (Fowler, Melnick & Mathieson, 1997).

Tremendous advancements in medicine, especially the development of combination drug treatments, have turned HIV/AIDS into a chronic illness for many women. The psychosocial impact of HIV/AIDS on women remains formidable, albeit always changing. Women who have contracted this disease face many uncertainties. They are living with a chronic illness frequently characterized by vague symptoms, rapidly changing medical management or philosophies and possible social isolation (Adair & Burian, 1997). Many of the women living with HIV in the United States are also mothers. These women are faced with the challenge of coping with their own illness as well as caring for family members (Cotton, 1999). The functionality of women living with HIV/AIDS can be compromised by the stigma of their illness, isolation, denial, concealment of their HIV status and difficulty accessing health care (Matheny, Mehr & Brown, 1997). Yet, many of these women are long-term survivors, having had the infection for many years and successfully meeting these challenges.

The Impact of Managed Care on HIV/AIDS and Nursing

The range of health care services delivered to women living with HIV/AIDS is broad and complex. In addition to the now standard use of combination drug therapy, for example, the use of complementary therapies for HIV/AIDS continues to rise (Gillett, Pawluch & Cain, 2002). Perhaps the most significant factor in the organization of health care delivery to these women, however, is the factor that increasingly defines health care for Americans in general – *managed care*. HMOs and PPOs now cover approximately 90% of insured workers, over half of Medicaid recipients, and 16% of those in Medicare. Medicaid provides access to health care for over 53% of all adults with HIV disease and over 90% of all children living with HIV/AIDS (Terry, 2000).

Managed care has affected nurses' work in many different ways. Nurses find themselves providing less direct patient care at bedside, a troubling development for many nurses. The primary nurse's traditional roles as caregiver, educator, and provider of emotional support remain, but are joined by new and potentially stressful responsibilities involving collaboration and negotiation. On the positive side, managed care creates new opportunities for leadership and growth (Miller & Apker, 2002).

For HIV/AIDS nurses, a new opportunity is the ability to focus on the psycho-social aspects of living with HIV/AIDS. Women living with HIV/AIDS are faced with new problems such as the complicated interactions of medications, in addition to those of everyday life. Effective health care for these women must include more than teaching about medication side effects and focusing on lack of adherence to prescribed treatments. Health care workers, especially nurses, must focus intensely on the psychosocial factors that support survival and those factors that nurses can influence (Strawn, 1995). *Inner strength*, a phenomenon described by various disciplines involved with the care of individuals experiencing challenging life situations, has been identified as a factor of psychological health and spiritual well being, as well as a dynamic component of holistic health (Dingley, 1997). *Inner strength*, therefore, appears to be a promising concept for use in psychosocial intervention.

The purpose of this paper is to review and critique the concept of inner strength as it is discussed primarily in the nursing literature. We will then refine it by integrating ideas from the sociological concept of the *existential self*. After providing illustrations from an interview-based study of women living with HIV/AIDS, we will argue that this integration will make the concept of *inner strength* more useful for research on women living with HIV/AIDS and for the application of this concept to clinical work with these women.

THE CONCEPT OF INNER STRENGTH IN THE HEALTH CARE LITERATURE

The concept of *inner strength* clearly reflects nursing's traditional emphasis on understanding health and illness through the patient's perspective, while celebrating the patient's efforts and ability to fight – if not overcome – illness. Rose (1990) described *inner strength* as the holistic, health-focused phenomenon of psychological health in women. She utilized a phenomenological approach to examine inner strength in nine healthy women. Through unstructured interviews, she identified nine essential themes of inner strength that represented psychological health. She suggested that, since these themes constantly merged and overlapped with one another, none be interpreted alone. The nine themes are: *quintessencing*, or the most perfect embodiment of self, included recognizing, becoming, accepting and being one's true self; *centering*, a process of focusing and balancing outside events with inner self; *quiescencing*, being calm, quiet and at rest; *apprehending intrication*, or knowing and understanding that which is complicated; *introspecting*, or gaining awareness of self and one's own psychological processes; *using humor*; *interrelating*, a personal, intimate, open and reciprocal process; *having capacity*, the ability to heal, solve problems, stay present, face pain and recognize a lack of capacity; and *embracing vulnerability*, a process of accepting, acknowledging and integrating one's imperfections and humanness. Rose's model is useful because it focuses attention on the complex, multi-faceted dimensions of psychological health, by generating a holistic concept to account for them. Her model is of limited value to the present discussion, however, because its themes are derived from the experiences of healthy women whose *inner strength* is implicit – it is not because it is not being tested by the awesome reality of facing life-threatening illness.

Moloney (1995) utilized a phenomenological approach to elicit stories from twelve women over the age of 65 that exemplified being strong. Three constitutive patterns, each with distinct themes, emerged in the analysis. The themes were surviving (living with loss and putting it behind you); finding strength (drawing strength from others, feeling good about self); and, gathering memories . . . seeing patterns (the process of telling the stories themselves). Moloney's investigation suggests, to nurses and their female clients, the benefit of understanding strength from a woman's perspective. Furthermore, her work is relevant to the present study because she illustrates the power of storytelling itself, while arguing for the value of exploring the experience of *inner strength* in the lives of socially and culturally marginalized women.

Roux and Keyser (1994) investigated *inner strength* in eighteen women with breast cancer and found that their experiences were manifested in the following four major themes: Coming to Know; Strength Within of She Who Knows; Connection of She Who Knows; and Movement of She Who Knows. These themes were representative of stages through which the women passed following a diagnosis of breast cancer, i.e. facing and accepting the diagnosis; acceptance of accompanying pain that opens the passageway for her strong spirit; a passionate connection of her instinctive life with her true loves; and finally, a passage through wandering and searching to inner strength used for moving, harmonizing and facilitating a desired change.

Roux and Keyser's work elegantly portrays the journey of self that women with breast cancer must navigate. This research can be critiqued, however, for its assumption that all women with breast cancer travel through the same stages of coping with cancer in the same order. The empirical question remains whether *inner strength* is experienced developmentally or situationally.

Dingley (1997) utilized grounded theory to examine *inner strength* in women with coronary artery disease. She described a multidimensional and multidirectional process of growth in inner strength that encompassed the total being holistically. Dingley described five interrelated constructs in the process of growth in *inner strength* in women recovering with cardiac disease. The first was allowing for *nurturance*, which involved receiving and accepting psychosocial support. *Dwelling in a different place* involved a process of focusing self outside of illness. *Balancing the search* was representative of balancing and assimilating new understanding, meaning and direction in their lives. The fourth construct, *healing in the present*, represented the process of creating a new, normal self in the present. Finally, *connecting with the future* occurred when women extended themselves to accept support from others. Dingley suggested that these five constructs could be identified in women with HIV/AIDS.

Related Concepts in Nursing

There are three related concepts in nursing that should be mentioned in this discussion. The first concept is *hardiness*, which is viewed as an attitude, personality characteristic and psychological construct that serves as a resource for resisting the negative effects of stress and enhancing long-term survival. Carson (1993) conducted a questionnaire-based, clinical study of women with HIV/AIDS. She found a positive relationship between hardiness and long-term survival, and perception of health, use of prayer and meditation, and healthy habits such as exercise and special diets. She concluded that a positive perception of one's health was associated with hardiness. Strawn (1995) explored *hardiness* with a review of scientific and anecdotal reports in the alternative health care literature to explore the role of nursing in long-term survivors of people with HIV/AIDS. She reported that possible factors associated with long-term survival are individual immune response, genetic differences, psychosocial factors such as internal resources and social affiliations, and differences in coping.

The second concept is *transcendence*, which refers to the ability to move beyond one's immediate predicament to a psychic or psychological space where hope resides. Coward (1995) based her work on self-transcendence on theories of Viktor Frankl (1969). Her survey-based investigation described the lived experience of self-transcendence in ten women with AIDS. She defined self-transcendence as "reaching out beyond the boundaries of the self to achieve broader perspectives and behaviors that help one discover or make meaning" (Coward, 1995, p. 314). Tsevat et al. (1999) surveyed fifty-one persons with HIV/AIDS regarding health status, quality of life, life satisfaction and attitudes toward family and friends; 71% of respondents indicated they were delighted, pleased or satisfied with their lives. Further results of the survey indicated factors other than health were found to be more important to a sense of satisfaction and a will to live. Improved living conditions, drug and alcohol treatment and addressing psychological and spiritual issues appeared to have an equal or greater effect on quality of life than medical care alone (see also Katz, 1996).

The third concept is *resilience*, which Hunter and Chandler (1999) described as an internal resource that allows one to be able to overcome adversity through use of developed internal and external protective mechanisms. Moreno and Simoni (1997) studied the relationship between stress and resilience among Latina women with HIV. Coping mechanisms utilized by these women were spirituality, non-traditional healing cures, therapeutic support groups that addressed their culture, gender and current situation and the presence of significant others in their lives. Zorrilla et al. (1996) designed a pilot clinical project to encourage resiliency by empowerment of women living with HIV. Strategies of self-help groups, a two-day seminar and part-time compensated work were utilized. Andrews (1995) described social support as a buffer to stress in her article focused on urban mothers with HIV. Within the context of poverty, HIV was seen as a stressor because of the stigma and the challenge it presents to individual psychological coping mechanisms. She concluded that survival for these particular women was possible because of help from their social support networks and strength from within themselves. Surviving incorporated managing daily life, raising children, dealing with spousal illness and caring for elderly parents. The description of help and strength from within the women was similar to the concept of inner strength and their social support served as a source for that strength.

All three concepts – *hardiness*, *transcendence*, and *resilience* – are psychologistic constructs. They focus on ideal coping resources and strategies that can hopefully be enhanced if not achieved through clinical intervention. Further, research on these concepts is typically conducted through surveys. In contrast, we will develop *inner strength* as a "socio-nursing" concept based upon the women's own ways of talking about their practical efforts to cope successfully in everyday life.

REFINING THE CONCEPT OF INNER STRENGTH

The literature on *inner strength* marks an important development in health care delivery in general and in nursing specifically. This concept includes several analytically positive characteristics. *Inner strength's* focus is on the subjective experience of illness. This patient-orientation makes it conducive to nursing's mandate to work with illness in terms of the ways the patient experiences it. Furthermore, since *inner strength* was developed in terms of women's illnesses and their distinctive experiences of illness, it can help resolve the remnants of male-bias in bio-socio-psycho-medical research on HIV/AIDS. Although the concept of *inner strength* appears most frequently in the nursing literature, it rests upon an interdisciplinary foundation. This concept borrows ideas from poetry, medicine, psychology, spirituality and autobiography, as well as nursing – and is thus conducive to the philosophy of holistic care.

The concept of inner strength also contains several weak analytical characteristics that are likely artifacts of its allopathic lineage. *Inner strength's* ontological status is that of a thing-like phenomenon. The concept is described as if it were an entity or quality deep inside the psyche or self of the patient, to be discovered by the researcher in an almost diagnostic way. In a sense, this formulation of inner strength resembles the early literature on *quality of life*, which attempted to locate this amorphous experience in quantitative, measurable terms as if it were a disease (Ragsdale et al., 1992). At the other extreme, *inner strength* is described in the literature in overtly metaphoric and romanticized terms that do not reflect the mundane, everyday life of living with an increasingly chronic albeit serious illness. Thus, *inner strength* is generally thought of as synonymous with doing well. Evidence offered for *inner strength* is the heroic ability to overcome the debilitating aspects of HIV/AIDS. The literature does not explicitly allow for women whose experiences of *inner strength* are real to them, yet who are incapable of overcoming the awesomely debilitating power of HIV/AIDS.

Existential sociology provides several useful ideas for refining the concept of *inner strength*. Existential sociology is a fairly recent school of social thought that is the work of Jack Douglas and his students at the University of California at San Diego (Douglas & Johnson, 1977; Kotarba, 1979, 1984; Kotarba & Johnson, 2002). The focus of their writing is on the individual's experience of everyday social life. People travel through life constantly seeking meaning for events, problems, situations, and the self. Social rules are crucial to people's experience of everyday life. Social rules do not, however, cause people to act certain ways. Instead, people use rules as resources for solving practical problems. Social life is situational. Rules are invoked, interpreted, and applied to make sense of actual situations people face.

Feelings are the wellsprings of social life and activity. The goal of all human behavior is the accommodation or satisfaction of feelings. Rationality is secondary and often used to account for feelings-driven behavior. People account for or make sense of everyday life, both to themselves and others, through story telling.

The *existential self* is the primary concept in this school of thought. By definition, the *existential self refers to an individual's unique experience of being within the context of contemporary social conditions, an experience most notably marked by an incessant sense of becoming and an active participation in social change* (Kotarba, 1984, p. 225). The self serves as the integration of cognitive (psychological), affective (embodied), and evaluative (social) responses to everyday life. People find themselves in a world not of their own choosing, but one to which they are forced to respond in order to manage everyday life and its problems, like illness.

Thus, existential sociology posits the person as a social actor who faces and attempts to master everyday life dilemmas. The person seeks meanings (e.g. values, rules, definition, and attitudes) from others – to complement deep feelings – as primary resources for dealing with these dilemmas. The actor manages meanings by embedding them in stories that form the substance of social life. Therefore, we conceptualize inner strength as *the different ways women with serious illnesses experience and, subsequently, talk about the deepest, existential resources available to and used by them to manage severe threats to body and self.*

Thus, we see *inner strength* as a process of experiencing, conceptualizing and talking about survival rather than an actual (psychological, ethical, genetic or hormonal) substance that somehow contributes to survival in a causal sense. Conceptualizations like this risk appearing tautological. Are we arguing that all women who survive HIV/AIDS experience *inner strength* simply because they are alive? No, we are simply focusing on and describing those women who talk about their coping with HIV/AIDS in terms of internal features we call *inner strength*. Furthermore, we will argue that our conceptualization of *inner strength* promises to be of clinical use in helping other women with HIV/AIDS cope with their illness and its effects on their sense of self.

METHODS

An important contribution of existential sociology to the concept of *inner strength* is a "patient-centered" methodological orientation. Existential sociology is committed to the complete integration of the respondent into the research enterprise. The goal is less to pose questions (of interest to the researcher) to respondents, than to allow respondents to tell their own stories of their everyday lives. These stories refer to the description of respondents' everyday lives, largely in terms of their words, meanings, feelings, and interpretations (Richardson, 1997). Through a process of *creative interviewing* (Douglas, 1985), the researcher functions largely as a trained and empathetic listener (see also Kotarba, 1990). In our interviews, we asked the respondents to tell us what it is like living with HIV/AIDS. We told them we were especially interested in the serious problems they face in dealing with their illness, and the sources of help they seek and use to overcome those problems. When our respondents talked about cognitive, affective or evaluative resources deeply embedded in themselves, we knew they were discussing what we knew as *inner strength*.

We collected the data for this study through 19 in-depth, biographical interviews. The respondents were HIV+ women living in the Houston, Texas metropolitan area. We assembled our sample through the theoretical sampling strategy suggested by grounded theory (Strauss & Corbin, 1980). Grounded theory refers to an inductive method of social research. Research questions and sampling are concurrent processes that unfold as the study progresses and discoveries are made. We publicized our study and asked for HIV+ women to contact us if interested. In all cases, we interviewed the women at times and locations convenient to them. Each respondent was given $50 for her time and assistance with the study. The interviews were directed, audio-taped, and 60 minutes in average length. All interviews adhered to the guidelines for the protection of human subject issued by the University of Texas Health Sciences Center, Houston.

This study was structured as a team project (see Douglas, 1985). All four members of the team contributed to all aspects and phases of the study, including design, interviewing, analysis and the construction of reports of findings.

The interviews were structured to minimize clinical bias while eliciting stories as naturalistically as possible. We avoided asking questions that assumed that our respondents used or even understood the concept of *inner strength*. After some preliminary questioning to obtain identifying "face sheet" information, we elicited respondents' stories two ways. First, we asked them to simply talk about the problems and crises associated with having HIV/AIDS and the things they did to overcome these problems and crises. Later in the interviews, we defined the concept of *inner strength* and asked them if they had similar feelings and experiences. To enhance the naturalistic tone of the interviews, we asked our respondents to tell us their stories of living with HIV/AIDS chronologically as well as biographically.

At the conclusion of the interviews, our sample consisted of: privileged, middle-class HIV+ women who either work in HIV care or as HIV/AIDS activists ($N = 5$); lower and working-class HIV+ women located at a inner-city social services agency ($N = 10$); and HIV+ women with late to in-stage AIDS ($N = 4$). The total sample had the following characteristics: average age: 33.6 (range was 19–54 years); Anglos: 4; African-Americans: 12; and Latinas: 3.

Twelve women identified themselves as being single, separated or divorced, whereas thirteen had living children. Thirteen respondents had either private insurance or Medicaid, but six were truly indigent and received their care in government-funded clinics. Only one woman had not attended any high school or received her GED. One half of the sample had either attended or graduated from college. Only seven women stated they were totally unemployed. However, one respondent worked as a prostitute and two had temporary work at the time of the interview.

Years of being infected with HIV ranged from one to twelve, with the average time of infection being 5.6 years.

THREE STYLES OF TALKING ABOUT ONE'S INNER STRENGTH

Arthur Frank (1995) elegantly portrays ill people as *wounded storytellers*. When ill people tell their stories of suffering, of hope, and of the hard work it takes to survive, they become active participants in their illnesses – no longer merely recipients of care. Whether they tell their stories to family members, health care workers or social researchers, their stories help them make sense of their illnesses and to bond with their bodies. Their stories also help heal the listener.

Of particular relevance to the present study is the way sick people understand how their relationship to the (social) world impacts their experience of illness. The *existential self* refers to a person's deepest sense of individuality and existence, and the inherent need to make sense of both. Serious illness like HIV/AIDS poses a major threat to the integrity of the self. *Inner strength* refers to the specific process by which women talk about the resources they use to survive if not master existential crises like HIV/AIDS. In our interviews, respondents told of the creative ways they sought to enhance, reaffirm, locate, interpret, discover, and/or celebrate inner strength – yet, always in the (social) presence of others.

Frank (1995, p. 76) insists that the researcher can analyze respondents' stories by organizing them into "types." A story – or narrative type – "is the most general storyline that can be recognized underlying the plot and tensions of particular stories" (Frank, 1995, p. 75) (see also Mishler, 1999). The purpose of typing stories is not to impose the researcher's voice over that of the respondents, but to help the listener play closer attention to the stories and sort out narrative threads.

Our interviews produced three types of stories in terms of overall themes. *Faith stories* recounted the ways reliance upon a higher power (spiritual or religious) provided a sense of *inner strength. Character stories* recounted the ways women experienced *inner strength* as a resource available to them before as well as during their illness. *Uncertainty stories* recounted the ways women perceived their *inner strength* as problematic. In all three types, the women describe *inner strength* as an accomplishment or process requiring collaboration with others. This contrasts markedly with the idea of some sort of inherent, psychologistic phenomenon that previous conceptualizations of *inner strength* have posited.

In order to illustrate our model of *inner strength*, we present the following first-person synopses of three stories. These stories are representative of the themes that emerge from all nineteen respondents, and they are simply among the more interesting stories we heard. Respondents, however, often wove different types of stories together during the course of our conversations, as they described different stages in living with HIV/AIDS. This methodological complexity is to be expected, for as the concept of the *existential self* tells us, our understanding of problems and events changes as we find ourselves in constantly changing social and embodied situations (Kotarba & Johnson, 2002).[1]

Faith Stories: Barbara

I interviewed Barbara in one of the offices at "The Place." The Place is a neighborhood social service agency in the middle of Houston's impoverished Fourth

Ward. They provide numerous programs for people with chemical dependencies, HIV/AIDS, and legal problems.

Barbara is a petite, thirty-four year old African-American woman with a constant smile on her face. She is a client of The Place, and proudly states that this agency is helping her put her life together. Upon further reflection, Barbara clarifies this point: The Place is not helping Barbara get her life back together so much as helping her get it together for the first time ever.

The message is: Barbara derives her *inner strength* from others. They include her husband whom she met at The Place, and her church, an inner-city Catholic Church. Her *inner strength* has been evolving over the course of her life as a person with HIV/AIDS.

The major turning point in her life, or *epiphany* (cf. Denzin, 1984), came in 1992 when she was convicted of possession of illegal drugs. For many years, she engaged in prostitution and theft to support her drug abuse and other vices that are part of the Fourth Ward lifestyle. At the time of her arrest, she was an IV drug user, and "speed ball" (i.e. injecting a mixture of heroin and speed or cocaine) was her drug of choice.

While she was locked up in Harris County Jail awaiting transport to the Texas Department of Corrections (TDC) to serve her term, she was diagnosed with HIV during her admission physical examination:

> (I first learned I was HIV+) in 1992 when I was in the county jail. I took the test and two weeks later they told me I was positive. I wasn't too surprised. I been shooting up with a woman and a man, and when I heard they died from AIDS, it wasn't really a surprise to me when I got it, so I didn't take it too hard.

When she was transferred to TDC, she tried to keep her HIV infection a secret. She was afraid that the other inmates would make fun of her, avoid her, and persecute her in other ways. Eventually, she was transferred to the medical unit.

Barbara says that it was very difficult being HIV positive while locked up in prison. There were times she considered suicide because her life was so "dysfunctional." The "Embrace the Cross Ministries" rescued her. This ministry works exclusively with women in prison. Through them, Barbara learned how to pray, first for the penal system so it will understand the problems women with HIV face, and second for help with her own problems:

> I also prayed for strength for myself. I prayed for strength not to turn bitter about it. Like the two ladies (in the prison who were also infected), you know, who felt everybody owed them something . . . like, someone else was to blame for their troubles. . . . I also prayed for strength to make it.

When she was paroled, she went to The Place for peer support, after-care services, and so forth. She went to The Place because she wanted to escape the bad friends and influences that contributed to her drug problems. Through The Place, she found her current job as a part-time warehouse worker.

Her husband is not HIV positive, but he is a recovering alcoholic and drug abuser. He is her second major source of strength. For example, she often gets tired of taking all the pills prescribed by her doctors at the local county hospital. Her husband, however, organizes her medications and encourages her to keep up with them. They live in an apartment located for her by a counselor at The Place.

Barbara said she had the possibility of *inner strength* before she contracted HIV. She insists that she could not take advantage of and nurture her *inner strength*. She ignored her faith in God when she was living on the streets.

Barbara defines *inner strength* as "being blessed . . . with the grace of God." She knows when she has it because she is able to stay alive and survive. But, *inner strength* by itself is not enough to survive. She has learned from all the therapy groups she has attended that one also needs high self-esteem. She can keep her self-esteem high by being able to talk about her HIV in front of a group. "Learning how to talk about it made me feel that being sick is not a bad thing anymore."

Barbara says that *inner strength* can help you be a good person when you are trying to be bad: "It can help you sort out and clear up things. I knew I was a good person when I was using and selling drugs, but I also saw myself as a bad person when I realized I was helping to spread drug use."

Inner strength can also help with practical problems. For example, *inner strength* (i.e. faith in God) helped her remain patient while she faced the very lengthy and difficult task of locating an apartment with a record as a convicted felon. "Having a felony is worse than having an illness."

Inner strength is a thing-like experience. "Either you have it or you don't." It does not vary in more-or-less terms, but you can lose it quickly. "Evil spells" are a major threat to her *Inner Strength*. Although Barbara always felt she had *inner strength* in her through the grace of God, she needed her family to bring it out and make it available to her:

> Women who don't have control over their HIV don't have family support. I have it now since I told my family about it (HIV). I made up with my aunt. She prays for me, she helps me know that God always loved me, but I had to do something about it myself.

Character Stories: Mary

When I first saw Mary Ellis, I wasn't quite sure it was she. We made an appointment to meet and talk at the Body Positive, a social agency that offers physical therapy to persons living with HV/AIDS. Mary agreed to an interview for the WISH (Women with Inner Strength) Project. I saw a diminutive lady walking down the hallway at BP, who did not appear to match the energy, pace and volume I heard over the telephone.

In metaphorical terms, Mary is an operator in front of a console. Her gaze shifts from meter to meter as she assesses data and readings. Her hands move from dial to dial as she adjusts energy and resources. But Mary is not in control of a factory or a transportation system. She is in charge of her self. And her self consists overwhelmingly of everything about her HIV/AIDS.

Since she was diagnosed in 1991, Mary has increasingly dedicated her life to mastering the illness. She contracted HIV from her boyfriend. She did not realize she was possibly infected until he got rapidly sick and died quickly from the disease he secretly passed on to Mary. She took care of him, as she continues to take care of others with HIV. Thus, Mary's commitment to HIV is not ego-centered. She realizes she is one of many. Her suffering is never "that bad," because she always compares herself to others worse off.

Mary has freed herself to work with HIV/AIDS by taking care of the logistics of her life. She lives on Social Security Disability, and receives her medical care through a Medicare-sponsored HMO.

Part of Mary's commitment to – perhaps obsession with – HIV/AIDS is related to her distrust of medicine. She consults several doctors, but does not let herself become dependent on any one of them:

> I'm afraid of the strong drugs doctors prescribe for AIDS, drugs that will kill you way before the virus gets you. You can't trust doctors because they only see things one, narrow way. They do not keep up with other ways of treating AIDS.

Yet, she takes all the prescribed drugs from all her doctors religiously. She insists that she maintains ultimate control over decisions regarding which drugs she will take, but she ends up taking a lot of them.

HIV/AIDS is not simply an illness that has invaded Mary's body. It has become the central theme of her self. Meaning for HIV/AIDS is meaning for life. Therefore, making sense of HIV/AIDS allows her to make sense of life. Accordingly, Mary seeks meaning for the puzzle of HIV/AIDS from as many sources as possible, relying on her better judgment to assess the veracity and usefulness of these meanings.

Mary's search for meaning for HIV/AIDS is overwhelmingly rational and intellectual. As existential sociology reminds us, however, the feelings and affective needs that permeate everyday life experience cannot simply be ignored or overridden through social definition (Kotarba, 1984). The Internet itself cannot make one feel warm and fuzzy. Mary realizes she needs a relationship in her life to close the circle:

> I need a man. I never thought I would ever say that, even before I got infected, but I'm tired of being alone. I'm tired of fighting my HIV by myself. I get tired. I don't care whether he's a

> man with HIV, or a man who will love me regardless of the fact I have HIV. I'm not simply a person with HIV. I'm more than that, or at least I want to be more than that. I want a companion I can have another life with.

Potential sources for a companion include a cruise, group activities or even the Internet!

Uncertainty Stories: Elaine

Elaine is a twenty-eight year old African-American woman. What is distinguishable about Elaine's appearance is the fact that there is nothing very distinguishable about her. She does, however, look fifteen or twenty years older than she really is. The wrinkled complexion, unkempt hair, and dirty black jeans do not denote a street prostitute, which is Elaine's professed occupation. Elaine does not fit the visual stereotype of the street prostitute – she simply looks poor, which she certainly is.

Elaine's goal in life is survival, survival on a day-to-day basis. In clinical terms, Elaine has a "dual diagnosis," that is, HIV/AIDS and chemical dependency. The latter problem is so overpowering, however, that Elaine's crack habit pretty much runs her life. She has little time, resources, or energy to care about or take care of her HIV/AIDS.

Surviving her crack habit is so overwhelming that, on most days, she does not have much time for anything else:

> I just don't have time to worry about my HIV. When I get sick, and I think it's from my HIV, I go down to LBJ (county hospital) and they give me some cocktail. I stop taking my medicine when it makes me feel bad. I feel bad enough from my habit; I don't need no more bad feeling.

There is some truth in Elaine's concern over her medicine. Adherence to highly active antiretroviral therapy (HAART) is very problematic among women who are depressed or who are substance abusers (Chesney et al., 1999).

Elaine's HIV/AIDS is not the only thing drowned out by crack. She has two children (two boys, ten and six) who live with their grandmother. Although Elaine rarely sees them, she is relieved that they are with family, not the county: "My mother takes good care of my children. She wants to adopt them, so the county can never take them away from us. I'll do it, if we get to keep them."

Elaine's shabby appearance means she does not have the motivation to make a good living from prostitution. She will perform sex for either money or, more likely, a rock. Not just a dime rock, but even just the promise of a chip. She hangs with men all day, the men on the corner who will tip her a chip for running errands. The men, White and Black, who pass by in cars pay her $25 for a trick, $50 with no condom. Some men make her freak (i.e. do any and all sorts of sexual behaviors) merely for the often unfulfilled promise of a rock.

Elaine's sense of *inner strength* is very different from that of most respondents. She said that my description of *inner strength* sounds like something that would be inside her, something that's a part of her. Elaine only believes in something that comes to her from the outside, something that is part of the day-to-day situation – luck:

> Either I'm lucky or I'm not. There's very little I can do to control my luck. All day long, you get ripped off . . . by dealers and pimps and men. . . . I know I got good luck when I can get my drug, not get beat up, and get through another day.

Taking care of her HIV/AIDS is not a choice Elaine makes. She cannot do much of anything else until and unless "she is cool with her habit." Her commonsense theory of luck states that she cannot do much to increase her luck, but she can do things to lose any good luck she has:

> You can lose your good luck by actin' a fool . . . by being stupid. Like, when you let a man run you. You gotta be smart on the streets because there are too many things that can go wrong and do you in.

Elaine's chemical dependency makes her totally self-centered. Luck not only explains her day-to-day success at survival, but it also explains other people's fate:

> I won't do anything bad to people, to do harm to others. Now, sometimes you let a man be with you without a condom, you know what I mean? If he's foolish to want no condom, then why should I worry about it? Whether he gets it (HIV) is up to him. I can't control his luck.

Elaine is not very skilled at telling stories about herself and her ability to cope with HIV/AIDS. Her world is not very "social," in the commonsense meaning of that word. Her everyday interactions with others – tricks, drug dealers, and public health nurses – are cold and instrumental. She does not frequently visit with the people who could possibly serve as listeners to stories about living with HIV/AIDS: her mother and her children. She verges on the brink of the ultimate existential dilemma: living a meaningless life alone:

> Does anybody help me with my HIV? I don't know . . . I don't think anybody knows about my HIV.

DISCUSSION AND CONCLUSION

These three stories illustrate the value of a concept of *inner strength* informed by existential sociology. We can now see women living with HIV/AIDS as active participants in mastering their dilemmas. They seek meaning for their otherwise absurd illness simply because, like everyone else in society, they cannot live for long in a state of meaninglessness. The sources of meaning for HIV/AIDS are socially distributed, that is, they are generally present in the women's everyday life world. As we have seen, these sources can range from church and social service agencies to a longed-for significant other and the Internet. These external meaning resources complement internal feelings that drive the management of everyday life. When their serious illness becomes focus of their everyday lives, women with HIV/AIDS develop great skill in fashioning stories about their sad, often heroic, but always complex experiences.

The social context of *inner strength* cannot be overemphasized. Spirituality may be experienced very privately and personally, yet, the value of spirituality and the ways it is applied to one's illness are made sense of in conversation with others. Although character stories recount the ways women experienced *inner strength* as a resource available to them before as well as during their illness, inherent *inner strength* is discovered and nurtured in the presence of others. Uncertainty stories, through which women talk about their *inner strength* being problematic, describe a sad and dysfunctional everyday life social world that does not make, contain or allow for the feeling of mastery that *inner strength* provides.

There are several important aspects of the experience of *inner strength* that we are unable to discuss in detail in this report of findings. For example, the women's experiences and sources of *inner strength* are likely to change over the course of an illness like HIV/AIDS. The illness itself not only changes, but the responses of the social world to and the women's ability to makes sense of (i.e. talk about) their illness also change. Of course, the reader should accept these findings and suggestions as tentative. They are derived from and based on a small, albeit fairly representative sample, given the strategy of theoretical sampling that guided this study.

Nevertheless, our formulation of the concept of *inner strength* contributes to the understanding of the everyday life of women living with HIV/AIDS by illustrating how the search for meaning for this serious illness is pervasive, intentional, mundane, and practical. This line of research tempers the aura of heroism present in the notion of *resilience* (Hunter & Chandler, 1999) and the original formulations of *inner strength*; of immalleable personality in the notion of *hardiness* (Carson, 1993); and of mystical spiritualism in the notion of *transcendence* (Coward, 1995). The concept of *inner strength* allows us to see women's search for meaning and control of illness in observable ways that can be appreciated humanistically yet enhanced clinically.

Implications for the Health Care Delivery System

Of what use is our investigation of *inner strength* for the delivery of managed care to women living with HIV/AIDS? Gareth Williams' (1984) notion of *narrative reconstruction* suggests an answer. Williams argues that open-ended interviews with ill persons provide them with the occasion to recall forgotten aspects of the illness, to reorder events to have them make more sense, and to explain their illness in ways more useful to the patient. Thus, the interview itself becomes therapeutic. The lesson here is that conversational interviews like those in our study enable ill persons to stop for a moment, map out their illness experiences, and see the big picture. As clinicians, we are reminded of the value of being empathetic listeners to our respondents' stories.[2]

The analytical typing of women's *inner strength* stories can pinpoint areas of intervention by health care workers. During episodes of physical or emotional crisis, clinicians may suggest prayer or supportive clergy to women who make sense of their ability to cope with HIV/AIDS through *faith stories*. Clinicians may also suggest new volunteer or social activities to women who see themselves thriving through *character stories*. Finally, women who describe their experiences of *inner strength* as extremely problematic may be telling *uncertainty stories* as a way of alerting clinicians to very serious issues related to surviving HIV/AIDS.

These suggestions for improved care all require fairly intensive interaction with patients in order to generate meaningful dialogue. This intensive interaction is part of and in fact called for by the philosophy of managed care. For example, a new role for nurses, that of *care coordinator*, has evolved. Care coordinators are registered nurses "charged with facilitating care delivery of patients' medical services from admission to discharge" (Miller & Apker, 2002). The role of care coordinator has remained ambiguous, giving the nurse great leeway in defining it. This professional autonomy creates a space for the serious clinical application of concepts like *inner strength*.

NOTES

1. All names and locations cited in this paper are pseudonyms.
2. See Haile et al. (2002) for a detailed discussion of the implications of this research for nursing care.

REFERENCES

Adair, M. L., & Burian, P. (1997). Evaluation of women. In: R. Muma, B. Lyons, Borucki & R. Pollard (Eds), *HIV Manual for Health Care Professionals*. Stamford: Appleton & Lange.

Andrews, S. (1995). Social support as a stress buffer among human immunodeficiency virus-seropositive urban mother. *Holistic Nursing Practice*, *10*, 36–43.

Carson, V. B. (1993). Prayer, meditation, exercise, and special diets: Behaviors of the hardy person with HIV/AIDS. *Journal of the Association of Nurses in AIDS Care*, *4*, 18–28.

Chesney, M. J., Ickovics, F. M., Hecht, G., Sikipa, J., & Rabkin (1999). Adherence: A necessity for successful HIV combination therapy. *AIDS*, *13*, 1–8.

Cotton, D. J. (1999). AIDS in women. In: T. C. Merrigan, J. G. Bartlett & D. Bolognesi (Eds), *Textbook of AIDS Medicine* (2nd ed.). Baltimore: Williams & Wilkins.

Coward, D. (1995). The lived experience of self-transcendence in women with AIDS. *Journal of Obstetric Gynecologic, and Neonatal Nursing*, *24*, 314–318.

Denzin, N. (1984). *On understanding emotion*. San Francisco: Jossey-Bass.

Dingley, C. E. (1997). *Inner strength in women recovering from coronary artery disease*. Unpublished master's thesis, Midwestern State University.

Douglas, J. D. (1985). *Creative interviewing*. Beverly Hills, CA: Sage.

Douglas, J. D., & Johnson, J. M. (Eds) (1977). *Existential sociology*. New York: Cambridge University Press.

Fowler, M. G., Melnick, S. L., & Mathieson, B. J. (1997). Women and HIV: Epidemiology and global overview. *Obstetrics and Gynecology Clinics*, *24*, 705–729.

Frank, A. W. (1995). *The wounded storyteller*. Chicago, IL: University of Chicago Press.

Frankl, V. (1969). *The will to meaning: Foundations and applications of logotherapy*. New York: New American Library.

Gillett, J., Pawluch, D., & Cain, R. (2002). How people with HIV/AIDS manage and assess their use of complementary therapies: A qualitative analysis. *JANAC*, *13*, 17–27.

Haile, B., Landrum, P. A., Kotarba, J. A., & Trimble, D. (2002). Inner strength among HIV-infected women: Nurses can make a difference. *JANAC*, *13*, 74–80.

HIV/AIDS Surveillance Report (2002, December). Atlanta: U.S. Department of Health and Human Services.

Hunter, A. J., & Chandler, G. E. (1999). Adolescent resilience. *Image: Journal of Nursing Scholarship*, *31*(3), 243–247.

Katz, A. (1996). Gaining a new perspective on life as a consequence of uncertainty in HIV infection. *Journal of the Association of Nurses in AIDS Care*, *7*, 51–60.

Kotarba, J. A. (1979). Existential sociology. In: S. G. McNall (Ed.), *Theoretical Perspectives in Sociology*. New York: St. Martin's Press.

Kotarba, J. A. (1984). A synthesis: The existential self in society. In: J. A. Kotarba & A. Fontana (Eds), *The Existential Self in Society*. Chicago, IL: University of Chicago Press.

Kotarba, J. A. (1990). Ethnography and AIDS: Returning to the streets. *Journal of Contemporary Ethnography*, *19*, 259–270.

Kotarba, J. A., & Johnson, J. M. (Eds) (2002). *Postmodern existential sociology*. Walnut Creek, CA: Alta Mira.

Matheny, S. C., Mehr, L. M., & Brown, G. (1997). Caregivers and HIV infection. *Primary Care Clinics in Office Practice*, *24*, 677–690.

Miller, K. I., & Apker, J. (2002). On the front lines of managed care: Professional changes and communicative dilemmas of hospital nurses. *Nursing Outlook*, *50*, 154–159.

Mishler, E. G. (1999). *Storylines*. Cambridge, MA: Harvard University Press.

Moloney, M. (1995). A Heideggerian hermeneutical analysis of older women's stories of being strong. *Image: Journal of Nursing Scholarship*, *27*, 104–109.

Moreno, C. L., & Simoni, J. M. (1997). Latina women and HIV: Coping and resilience (online). *National Conference on Women and HIV*, 168, Abstract No. P1.51. Abstract from: NIH/NLM AIDSLINE. Citation ID 97927468.

Ragsdale, D., Kotarba, J. A., & Morrow, J. R. (1992). Quality of life of hospitalized persons with AIDS. *Image*, *24*, 259–265.

Richardson, L. (1997). *Field of play*. New Brunswick, NJ: Rutgers University Press.

Rose, C. F. (1990). Psychological health of women: A phenomenological study of women's inner strength. *Advances in Nursing Science*, *12*, 56–70.

Roux, G. M., & Keyser, P. K. (1994). Inner strength in women with breast cancer. *Illness, Crises and Loss*, *4*, 1–11.

Strauss, A. L., & Corbin, J. (1980). *The basic of qualitative research* (2nd ed.). Thousand Oaks, CA: Sage.

Strawn, J. M. (1995). Long-term survivors: Thriving with human immunodeficiency virus/acquired immunodeficiency syndrome. *Holistic Nursing Practice*, *10*, 29–35.

Terry, K. (2000). Where's managed care headed? *Medical Economics* (April 10th), 1–13.

Tsevat, J., Sherman, S. N., McElwee, J. A., Mandel, K. L., Simbart, L. A., Sonnenberg, F. A., & Fowler, F. A. (1999). The will to live among HIV-infected patients. *Annals of Internal Medicine*, *131*, 194–198.

Williams, G. (1984). The genesis of chronic illness: Narrative re-construction. *Sociology of Health and Illness*, *6*(2), 175–200.

Wofsy, C. B. (1995). Gender-specific issues in HIV disease. In: M. A. Sande & P. A. Volberding (Eds), *The Medical Management of AIDS*. Philadelphia: W. B. Saunders.

Zorrilla, C., Santiago, L., Huertas, J., Carreras, A., Schute, P., Cintron, L., Martinez, J., & Pacheco, E. (1996). The role of empowerment as a health care strategy for women with HIV: Encouraging resiliency (online). *International Conference on AIDS*, *11*, 344. Abstract No. Tu. C.2460. Abstract from: NIH/NLM AIDSLINE. Citation ID 96922653.

PARADIGM TENSION IN MANAGEMENT OF CHRONIC DISEASE

Nancy G. Kutner

ABSTRACT

Health promotion and rehabilitation models of care are valuable for persons with chronic health conditions, but when these individuals are dependent on a life-maintaining technology, such as kidney dialysis, a cure-oriented model may dominate the system within which they receive care. Providers can preserve their monopoly over expert treatment knowledge by defining the key care issues, by limiting patients' access to expert knowledge, and by discrediting the patient as a responsible actor. Multiple care paradigms can benefit patients with chronic conditions, however, empowering the patient-actor to collaborate with the clinician to maximize functioning and well-being as well as patient survival.

INTRODUCTION

The majority of patients receiving health care in the United States have chronic, disabling conditions, and a curative medical model is unlikely to address all of the health care needs of these individuals. At the same time, patients who have no reversible component to their disease constitute only a small subset of those with

Reorganizing Health Care Delivery Systems: Problems of Managed Care and Other Models of Health Care Delivery
Research in the Sociology of Health Care, Volume 21, 107–123

ISSN: 0275-4959/doi:10.1016/S0275-4959(03)21006-1

chronic conditions. Thus, many individuals who have chronic health conditions can benefit from cure-oriented medical treatment, but they also can benefit from care that has goals other than cure. Combining different models in patient care is challenging, however, because each model has its own inherent assumptions, attitudes, and values, i.e. the different models represent different paradigms. The curative model of medical care continues to be the dominant model in this country, a situation that has been labeled a "residual problem" in our system of health care (Fox, 1997).

This paper uses the case of care for chronic kidney disease patients on dialysis to illustrate paradigm tension between the cure-oriented model of medicine and other relevant models, especially health promotion and rehabilitation models. Dialysis patients' dependence on technology for their survival reinforces the predominance of the cure-oriented model. At the same time, health promotion and rehabilitation models can make important contributions to dialysis patient outcomes. Like other persons with (and without) a chronic illness, better informed patients are more likely to become involved in health promotion behaviors, and the more that individuals are willing to actively pursue strategies to improve their health status, the better their health outcomes are likely to be.

An important difference between the cure paradigm and health promotion or rehabilitation paradigms in medical care is that these paradigms are centered on different responsible agents. The physician is the agent credited with achieving cure, but the patient actor must be ultimately responsible for health promotion and rehabilitation achievements. Clinicians, understandably, want to preserve their role as the owners of expert knowledge about cure-focused, or at least symptom-managing, treatment. Clinician monopoly over expert treatment knowledge is challenged by a patient-actor who pursues health promotion and rehabilitation goals as a partner in the care system.

Cure-oriented medicine is interested in patients as organ systems, not as thinking, feeling actors (Fox, 1997). This is part of the inherent set of assumptions, attitudes and values associated with the curative model of medical care. Cure-oriented medicine tends to fractionate human beings into molecules, cells, and organ systems – the "facts" of hard medical science. When the primary goal is cure, these facts, rather than the patient as a responsible agent, are central. This perspective reinforces the predominance of the curative model of care along with the expert knowledge of those who design and deliver life-saving technology.

The case of chronic dialysis care is a particularly useful one to illustrate why predominance of the curative model tends to be a residual problem in medical care. Before turning to characteristics of the dialysis care system, differences between medical care models are outlined. The curative model is contrasted with health promotion and rehabilitation models of care, and the applicability of multiple models for care of patients with chronic conditions is discussed.

MODELS OF MEDICAL CARE: WHERE DOES CHRONIC DISEASE MANAGEMENT FIT?

Cure, one of the most important goals of medicine, in its most complete sense can be defined as "the eradication of the cause of an illness or disease" (Pellegrino & Thomasma, 1997). Other appropriate goals of medicine include promoting health, preventing illness and injury, restoring functional capacity, avoiding premature death, and relieving suffering (palliative care). These goals overlap and can be considered as varying emphases along a care continuum, with cure and palliative care anchoring the two ends of the continuum (Fox, 1997).

In its purest form, the curative model concentrates on the goal of cure and does not consider the other goals of medicine. The approach is distinctly analytic and rationalistic. Clinical concerns are approached as puzzles to be solved, and patient encounters become occasions for scientific inquiry. According to this model, effective cure is contingent on effective diagnosis and treatment. Thus, the cure-oriented approach is highly invested in a scientific, biomedical approach to clinical practice. Treatments that are considered most effective are ones that improve objective, disease-related outcomes, especially survival. In a cure-oriented model, treatments that improve subjective and non-specific outcomes, e.g. patient-assessed quality of life, have low priority (Fox, 1997).

A model quite opposite to the curative model is the palliative care model. Relief of suffering, control of symptoms, and maintenance of functional capacity are legitimate goals of the palliative model, but cure and prolonging death are not legitimate goals. Palliative care considers the patient's subjective experience of illness to be as important as objective clinical data. A treatment is only considered appropriate if it is worthwhile from the patient's perspective. Finally, the management plan is tailored specifically to each patient in the palliative care model. This contrasts markedly with the curative model, in which the determination of appropriate treatment is viewed as a scientific question decided by existing medical knowledge and established clinical practices.

In reality, neither the purely curative model nor the purely palliative model is likely to be suitable for most patients, and these are not the only available models. As noted above, medicine has several other legitimate goals, including promoting health and restoring functional capacity, and these goals can be incorporated in various combinations into the care for a specific patient. Management of patients with a chronic or disabling health condition is facilitated by a multi-goal approach, with different emphases placed on specific goals at various stages in the chronic care process.

When patients require a period of aggressive intervention to be stabilized in the course of a chronic disorder, this is typically followed by a prescribed course of rehabilitation therapy. Examples include care for persons who incur traumatic brain injury, stroke, heart attack, and other cardiac events that require an interventional procedure. In these disorders, progression from reliance on the curative model to reliance on a rehabilitation model is an institutionalized part of the care pattern.

However, when the patient requires an *ongoing* procedure-based intervention, as in chronic dialysis, the starting point for a course of rehabilitation is less clear. Rehabilitation may be considered relevant and desirable, but the ongoing technological intervention tends to take precedence. This is understandable when the patient remains highly unstable and continued survival is problematic. Chronic dialysis is for the most part a routine outpatient procedure, however. The overall health status and functioning of the majority of dialysis patients can benefit from health promotion and rehabilitation strategies as well as from clinical efforts to fine-tune the dialysis technology (Painter & Johansen, 1999). Paradigm tension exists in the dialysis care system because of the essential differences between the inherent assumptions, attitudes and values of the cure-oriented dialysis technology approach to care versus health promotion and rehabilitation models of care.

DIALYSIS CARE: TECHNOLOGY-BASED CURE VERSUS HEALTH PROMOTION AND REHABILITATION GOALS

Dialysis Technology

In patients diagnosed with chronic kidney failure, the kidneys no longer adequately filter out toxic compounds from the body and maintain the volume, composition, and distribution of body fluids essential for well-being. Patients require a "renal replacement therapy" in order to survive. Dialysis, which removes toxins and excess fluid from the patient's blood, and kidney transplantation are the two available types of renal replacement therapies. The majority of patients are maintained by chronic dialysis because there is a shortage of kidneys available for transplantation and because not all kidney failure patients are appropriate candidates for transplantation. Of the approximately 379,000 patients receiving treatment for kidney failure in the U.S. at the end of calendar year 2000, 72% were on chronic dialysis therapy (USRDS, 2002).

The most widely used dialysis procedure is hemodialysis, in which the patient's blood is circulated outside the body through an artificial kidney machine. Kidney dialysis is considered a "halfway technology," because it provides imperfect replacement for normal kidney function. Improving the effectiveness of dialysis in removing toxins and minimizing uremia (symptoms such as lethargy and loss of appetite) is an ongoing effort (e.g. Hakim, 1992; Raj et al., 1999; Ronco et al., 1994; Twardowski, 2003).

This paper considers dialysis as a technologically-based intervention representative of the cure-oriented approach to medicine. The clinical delivery of dialysis is consistent with the characteristics of the curative model outlined above. It is important to note, however, that kidney dialysis is not a cure for chronic kidney disease and kidney failure, the patient's underlying disorder. Even the most efficient types of dialysis only provide 10–15% of the kidney's waste-removing functions, and they do not substitute for any of the multiple hormonal functions of the normal kidney. Dialysis produces a state of "controlled" uremia, and it is in this sense that dialysis can be considered to be consistent with a curative model of medicine.

Keeping patients alive and reducing their morbidity as much as possible have been the primary objectives of care since the introduction of dialysis technology. But whenever cure is not possible and patients are living with a chronic disease, maximizing patients' functioning and their quality of life should have high priority. Congressman Stark, a long-time supporter of Medicare funding for dialysis, has lamented that "little attention has been given to the patient in the policy debates over the End Stage Renal Disease (ESRD) Program. Reams have been written about the proper level of payment to renal facilities and physicians. Yet . . . we . . . do not have a good 'patient-outcome' or 'quality-of-care' standard" (Stark, 1993, p. 140).

Relevance of Health Promotion and Rehabilitation Models for Dialysis Care

Health promotion and rehabilitation are interrelated paradigms of health care that can make important contributions to the overall health and well-being of patients who have a chronic disease. *Health promotion* is used in this paper to refer to attitudes and behaviors that help to maintain and/or improve health status. *Rehabilitation* is used to mean a process that assists, or extends an opportunity to, individuals to regain ability to function optimally in their usual environment. Neither the health promotion nor the rehabilitation paradigm is well developed in the context of the actual delivery of dialysis care, however. This is especially curious with respect to rehabilitation, because rehabilitation is a concept that has been salient from the beginning of the Medicare program that provides treatment funding for the majority of dialysis patients in the U.S. Rehabilitation was in fact the "promise" on which the federal ESRD program was founded, as the following brief historical overview indicates.

When the ESRD Program of Medicare was initiated in 1973, members of Congress were convinced that the program would be cost effective because most of the patients restored to medical health via dialysis would be productively engaged, tax-paying citizens. Fewer than 4,000 persons were receiving dialysis in the U.S. at that time (Rettig, 1980). The treatment was expensive, and dialysis machines were not widely available. Patients selected to receive dialysis were considered good candidates for successful rehabilitation. Preference was given to individuals who did not have serious comorbid complications, who were relatively young, and who had secure employment or good employment prospects to which they could return (Evans et al., 1981). Thus, in the early years of the ESRD Program, *expectations* for dialysis patients' outcomes were quite optimistic. Senator Hartke, who proposed the ESRD Medicare legislation, predicted that 40% of patients treated by chronic dialysis could continue their existing jobs and that the remaining 60% would return to work following minimal "retraining" (Rettig, 1980). Full rehabilitation, especially vocational rehabilitation, was the expected outcome, with dialysis serving as a technological means to that end.

In the years following the initiation of the ESRD Program of Medicare, more and more dialysis facilities opened, and the criteria by which patients were admitted for treatment became less stringent. Patients' actual or potential employment status lost relevance as a consideration for receiving dialysis treatment. Patients with more complicated medical status, especially diabetics, were accepted for dialysis, along with older patients.

Surveys conducted since 1980 have repeatedly shown that no more than 20% of working-age dialysis patients are in the labor force (Curtin et al., 1996; Kutner et al., 1991). These data are not consistent with the early predictions that were made about patients' ability to continue or quickly return to employment after beginning dialysis therapy. Multiple factors contribute to low employment rates among dialysis patients, including patient socioeconomic characteristics and various labor force, employer, and disability insurance variables (Holley, 1994; Kutner et al., 1991; Porter, 1994). Patients' employment status is an especially visible marker for judging the success of a process of medical restoration, and a disappointing level of vocational rehabilitation is to some degree embarrassing for dialysis care providers, although vocational rehabilitation outcomes are influenced by factors beyond the clinical setting for which physicians have responsibility.

Capacity for physical activity (i.e. physical rehabilitation status), and mental health and emotional well-being (or psychosocial rehabilitation status), are also important rehabilitation dimensions. On health status surveys, dialysis patients score significantly below age and gender adjusted samples of the general population on scales assessing patient-rated physical ability to perform usual role activities, general health perceptions, energy/fatigue, and physical functioning (ability to perform activities such as stair climbing that require mobility, strength, and endurance). Dialysis patients' emotional well-being and social functioning, as measured by health status survey data, are closer to scores obtained for the general population. Multiple studies conducted in the U.S. (Diaz-Buxo et al., 2000; Hays et al., 1994; Kurtin et al., 1992; Kutner et al., 2000; Levin et al., 1993; Meyer et al., 1994; Walters et al., 2002) indicate that chronic dialysis patients' mean scores on the health constructs measured by the Medical Outcomes Study Short-Form 36, or MOS SF-36 (Ware & Sherbourne, 1992), have a very consistent rank ordering, with scores for physical role-functioning indicating that this is the most problematic, and scores for emotional well-being indicating that this is the least problematic, health status domain.

Dialysis patients' physical functioning can be influenced significantly by targeted rehabilitation efforts and by health promotion strategies that are similar to strategies advocated for the general population (Kutner et al., 1997). Exercise is a rehabilitation intervention capable of improving physical health and well-being in many disease populations; it is also a means of health promotion in the general population. Regular participation in physical activity/exercise is an especially good example of a non-dialysis strategy that is likely to have positive consequences for chronic dialysis patients' lives. Dialysis patients who participate in regular physical activity/exercise regimens show improvements in work capacity, lipid metabolism, glucose metabolism, and hypertension, as well as improvements in perceived mood and sense of well-being (e.g. Goldberg et al., 1979, 1983; Painter & Johansen, 1999; Painter & Moore, 1994; Painter & Zimmerman, 1986; Painter et al., 2000). In spite of these documented benefits and the fact that cardiac causes are the primary reason for death among dialysis patients, patients are not routinely instructed about exercise regimens and encouraged to be physically active (Painter, 1993).

Health promotion behaviors on the part of patients with chronic health conditions represent an effort to be as fully in control of one's health as possible. Individuals who are knowledgeable about their disease and about symptom management can become effective partners in the care process, which should ultimately benefit both the patient and the physician care-provider (Lundin, 1985). Promoting patient knowledge and involvement is not necessarily characteristic of dialysis facilities, however, as indicated by this patient's comments:

> From Day 1, you are kept in the dark . . . I don't know if some doctors never thought about it; . . . some may think that the patients can't understand it . . . You always have people that no matter what they know they are not going to use it, but that shouldn't stop you from educating those that will . . . I was led to believe that I could live indefinitely on dialysis. You look around

> and you start to notice people die. Through phases people drop like flies. My health was really going downhill, . . . and I basically had to go in and request my files . . . A patient that goes to my center . . . wanted to (change type of dialysis). He had to ask for his records to get it done. They refuse to talk to you about it. You don't have a choice if it is not the doctor's opinion . . . (Kutner, 1987, p. 59).

CONTROLLING THE EXPERT KNOWLEDGE BASE: CLINICIAN STRATEGIES

Defining the Key Issues of Care

Consensus conferences are an important professional setting for defining key issues in patient management. A conference titled "Strategies for Influencing Outcome in Pre-ESRD and ESRD Patients" sponsored by the National Institutes of Health (NIH), the American Society of Nephrology, the Renal Physicians' Association, and the National Kidney Foundation was held in Washington, D.C., in June 1998. An advance flyer advertising the conference gave this rationale for the meeting:

> The 1989 Morbidity, Mortality and Prescription of Dialysis Symposium, held in Dallas, was a major turning point in the evolution of dialysis therapy. That symposium brought attention to the importance of managing potentially correctable factors associated with an increased risk of death such as sub-optimal delivered dialysis dose . . . While improvements have been documented, the mortality of dialysis patients remains unacceptably high, indicating that there are other factors influencing patient outcomes that have not yet been sufficiently addressed . . .

Topics addressed in the 1998 conference included "Delivered Dose of Hemodialysis Therapy and Outcome," "Nutrition and Outcomes," "Anemia," "Cardiovascular Disease and Outcomes," and "Strategies for Preparing the Patient for ESRD." Health promotion and rehabilitation topics were not part of the agenda for this conference nor the agendas of subsequent NIH consensus conferences in the area of chronic kidney failure. A conference titled "Depression and Mental Disorders in Patients with Diabetes, Renal Disease, and Obesity/Eating Disorders," was held at NIH in 2001, but pharmacologic approaches to treatment tended to dominate the program.

The dialysis care system in the U.S. cannot fairly be described as illustrating only a cure-oriented approach to medical care. Individual clinicians and dialysis treatment centers have made efforts to combine state of the art dialysis technology with patient education and other health promotion and rehabilitation-focused programs (e.g. Curtin et al., 2002; McMurray et al., 2002; Medical Education Institute, 1994, 1995; Rasgon et al., 1993; Stivers, 1996). In addition, over the past decade a consortium of health professionals, researchers, and patients has collaborated to develop a wide variety of educational materials promoting rehabilitation strategies and with private support has made these materials available to dialysis clinic personnel, patients, and professional renal organizations. This consortium has also worked to influence policy agendas of government and renal organizations with the goal of promoting increased emphasis on rehabilitation of chronic kidney disease patients (www.lifeoptions.org). However, no financial incentives to encourage dialysis providers' achievement of rehabilitation goals have been established within the structure of the Medicare-funded program that oversees the system of dialysis care in the U.S.

Restricting Access to Expert Knowledge

Life after a diagnosis of chronic kidney failure means living with a number of uncertainties. Dialysis treatments give "borrowed time" (Chyatte, 1979), but how much time is unknown. Significant patient risks are associated with hemodialysis, peritoneal dialysis, and kidney transplantation. Persons with kidney failure face uncertainty about when and to what extent they will experience degenerative changes associated with their underlying disease. For example, without normal kidney function, a calcium/phosphate imbalance is created that contributes to metabolic bone disease. Neuropathies, restless leg syndrome, sexual dysfunction, and sleep disorders are common among dialysis patients.

Because the disease and treatment course can vary greatly among dialysis patients, individuals' awareness and monitoring of their own illness experience is crucial. In an effort to raise patients' awareness about core indicators of dialysis treatment adequacy, a brochure titled "*Know Your Numbers*" was prepared in 1995 with government support, for distribution to patients at all U.S. dialysis facilities. The brochure was designed to stress in simple language the importance of patients' understanding and monitoring dialysis treatment parameters. A report issued by the Office of the Inspector General (OIG) in March, 1997, however, indicated that the "*Know Your Numbers*" brochures received limited distribution within dialysis clinics. The OIG concluded that brochures actually reached only about one-third of the clinic dialysis patient population. Moreover, most of the patients who did receive brochures "were not familiar with the tests used by their dialysis facilities to measure adequacy of dialysis and the associated target numbers" (Neumann, 1997, p. 8). Among patients receiving the brochure, fewer than half reported receiving an explanation of the brochure's content or having it read to them by dialysis staff. Particularly interesting was the OIG's discovery of a correlation between a facility's distribution of the brochure and the frequency with which the facility routinely calculated adequacy measures of patients' delivered dialysis therapy.

Facilities that calculated the numbers on a monthly basis, rather than quarterly, were much more likely to also distribute the brochure to patients and to have patients who "participate more proactively in their dialysis" (Neumann, 1997, p. 10).

Sehgal et al. (1997) reported that many of the 145 dialysis patients they interviewed in Cleveland facilities were "poorly informed about the adequacy of their treatment as based on biochemical indices, the relationship between amount of dialysis and life expectancy, and the protein content of food items." Similarly, in our 1996–1997 interviews with 228 chronic dialysis patients in the Atlanta area, fewer than 20% of patients were aware of their own key indicators of treatment status (hematocrit level, serum albumin level, or values reflecting adequacy of delivered dialysis). These data suggest that patient education efforts are limited and/or not sufficiently effective.

Discrediting the Patient as a Responsible Actor

Non-compliance (or adherence) to treatment prescriptions is problematic in virtually all medical therapies (Sherman, 1996). It is a highly visible problem with dialysis patients because the markers that are used – missing a dialysis session, cutting short a dialysis session, excessive weight gain between dialysis treatments, and abnormal lab values that reflect diet non-compliance – are readily available to dialysis staff. Because the therapy places multiple demands on patients, expressions of non-compliance may represent patients' effort to take back some control over their lives (Curtin et al., 1997; Kutner, 2001). The large number of medications and diet and fluid restrictions that are required can burden patients who already have other lifestyle restrictions that chronic dialysis places on them, as well as burdens imposed by comorbidities such as diabetes.

A 1997 survey of over 2,000 dialysis patients indicated that the majority do not skip treatments or shorten treatments. The data also showed that patients desired more information about their treatment and increased involvement in decision making (Lore, 1997). However, several dialysis clinician-investigators have argued that failure of dialysis technology to achieve optimal outcomes is, to an important degree, a function of the patient actor's failure to be a faithful treatment complier (Kimmel et al., 1996; Kobrin et al., 1991; Rocco & Burkart, 1993). "Patient demand" is frequently cited as the reason for patients' receiving *delivered* dialysis treatment that is shorter than their *prescribed* length of treatment. Emphasizing non-compliance essentially "blames the victim," however, and deflects attention away from providers' failure to effectively communicate the importance of compliance, a health promotion strategy, for patients' own health and well-being.

WINDOWS OF OPPORTUNITY FOR AN EXPANDED DIALYSIS CARE PARADIGM

Several promising windows of opportunity have emerged over the past two decades for an expanded emphasis on rehabilitation throughout the dialysis care system. These windows have primarily been associated with technological innovations or changes in the delivery of technology. The health status outcomes "movement," which specifically identifies functional status and health-related quality of life as meaningful outcomes, can also be seen as a source of opportunity, however. I suggest that technology-based windows have not stimulated a significant increase in rehabilitation outcomes precisely because they are based on technological innovations that are inherently cure-oriented.

Technology as a Potential Catalyst for Rehabilitation

In the early 1980s an alternative to hemodialysis, continuous ambulatory peritoneal dialysis (CAPD), became available. CAPD can be performed by the patient at home and does not require use of a dialysis machine. This innovation was expected to facilitate patients' ability to continue employment and to maintain a normal lifestyle. However, fewer than 8% of dialysis patients in the U.S. were using CAPD as of the end of 2000. Our research indicates that CAPD patients have increased needs for support from family as well as from dialysis care-providers in order to continue performing their dialysis exchanges multiple times each day, every day of the week. Thus, important psychosocial issues must be addressed with CAPD patients (Kutner, 1998; Kutner et al., 1993). The clinical community has focused more on the physiological aspects of CAPD, its adequacy as a type of dialysis, and potential improvements in the technology of CAPD than on the associated psychosocial issues that this therapy creates for patients.

A second technological innovation that was widely viewed as promising for patients' potential rehabilitation was the development in the late 1980s of human recombinant erythropoietin (rHuEPO), a product of DNA technology, to improve the anemia that dialysis patients experience. Observed outcomes of rHuEPO therapy have included increased hematocrits and reduced need for blood transfusions and some improvement in exercise tolerance. Patients receiving rHuEPO are no more likely to maintain employment after beginning chronic dialysis, however.

Although it is not a newly developed technological innovation, the value of longer and more frequent hemodialysis treatments (compared to the conventional pattern of 3–4 hour treatments three times/week) began to receive increasing attention in the late 1990s. A decrease in uremic symptoms has been observed with more intense, or "daily," hemodialysis, but evidence to date is based on small numbers of patients in a small number of centers. Proponents are convinced that patients experience improved quality of life and are able to be more active, including being able to pursue employment (Mohr et al., 2001). There is limited clinical experience with patients' long-term acceptance of longer and more frequent hemodialysis and no evidence to date regarding outcomes associated with integrating this therapy regimen with an intervention such as an exercise protocol.

Health Status Outcomes: Physicians and Patients as Joint Experts on Medical Care?

Collection of a variety of outcome data, including patient assessed outcomes, and the use of these outcome data to inform the development of clinical practice guidelines, are hallmarks of the health services research and quality assurance initiatives. An important objective of two conferences organized by the Institute of Medicine in 1993 and 1994 was to incorporate patient assessed functional status measures into dialysis clinical practice and thereby stimulate increased attention to rehabilitation objectives in chronic dialysis care (Rettig, 1995; Sadler, 1996; Schrier et al., 1994). Although there may be persistent skepticism among medical professionals as to "whether the systematic acquisition of functional status data can improve on the intuitive judgment of clinicians" (Rettig, 1995, p. 199), the relevance of patient-assessed functioning and well-being for patient care as well as for evaluating outcomes of clinical studies is increasingly acknowledged (e.g. Bearon et al., 2000; Kaplan & Schenkman, 2000). The patient actor, as well as the professional health care provider, has "expert knowledge" in a system that incorporates patient-assessed health status outcomes as markers of quality of care. Ideally, patient assessments would help shape the care system in this paradigm, and health promotion and rehabilitation goals would be facilitated.

CONCLUSION

It has often been noted that chronic diseases and health conditions do not fit well within the classic medical model that emphasizes cure. However, cure is likely to be of much greater interest to a clinician than is teaching a patient how to be an informed manager of his/her own health and how to prevent secondary disability. Predominance of the curative model is a "residual problem" in American medicine, even when patients have a chronic condition for which cure, in the pure sense, is not possible.

When a sophisticated treatment technology is central to the survival of patients with a chronic disease, the tension is heightened between emphasis on a cure paradigm and emphasis on paradigms such as health promotion and functional restoration that are especially appropriate to chronic disease management. Providers hope that further refinements of technology will produce better disease control, and patients in turn often see technology as the primary determinant of their health status. The danger is that the promise of technoscience obscures attention to contributions that the patient as actor can make to his/her own health status.

Analyzing the kidney dialysis system of care in the U.S. as of 1980, Drees and Gallagher (1981) predicted that the importance of the rehabilitation, or restoration, paradigm would become increasingly acknowledged. They likened the emphasis on dialysis technology and the neglect of emphasis on the patient actor's involvement in the care process to a situation of cultural lag. They predicted that rehabilitation programs would emerge belatedly, in an effort to "catch up" to the technology and complement it.

As noted above, examples of effective health promotion and rehabilitation activities can be found in the dialysis care community. The paradigm blending that Drees and Gallagher forecast has not occurred, however. Windows of opportunity for expanding the dialysis care paradigm have been primarily technology-based windows. Technoscience and improvements in technology remain the central focus of dialysis providers, reflecting the strength of the curative model. This paper argues that the dialysis care system's limited attention to health promotion and rehabilitation goals has less to do with cultural lag between technological innovation and subsequent change in nonmaterial culture, as Drees and Gallagher (1981) suggested, than it does with tension between the priority of different care paradigms. Tension over ownership of expert knowledge can emerge when patient actors share the responsibility of disease management with clinician care providers. Unfortunately, blaming the victim for poor outcomes is one means of maintaining clinician ownership of expert knowledge.

Our interviews with dialysis patients indicate that "dependence on doctors and other medical staff," an item included in the "Kidney Disease Quality of Life" instrument (Hays et al., 1994), does not bother patients as much as feeling dependent on a machine bothers them. They want to be viewed as human beings interacting with other human beings rather than human beings whose lives depend on a machine. Clinicians who collect patient-assessed outcome data have reported gaining appreciation for patients' ability to contribute information that proves useful to the clinician in the care process (Meyer et al., 1994). However, in a technology-based system of care, the curative model is likely to remain strong. The extent to which dominance of the curative model will always be a residual problem in the dialysis care system, as opposed to a *merger* of paradigms relevant for chronic disease management, is a "perplexing . . . question about the role . . . of medical science, and of medical machines" (Peitzman, 1996, p. 280).

ACKNOWLEDGMENTS

Research support was provided by grant R01 DK42949 from the National Institutes of Health. The author is grateful for patients' participation in the research and for insights shared by colleagues in the dialysis care community, especially John Bower, Chuck Brown, Roberta Braun Curtin, Bruce Lublin, John Newmann, Wayne Nix, Edith Oberley, and Patricia Painter. An earlier version of this paper was presented at the 1998 annual meetings of the American Sociological Association.

REFERENCES

Bearon, L. B., Crowley, G. M., Chandler, J., Robbins, M. S., & Studenski, S. (2000). Personal functional goals: A new approach to assessing patient-centered outcomes. *Journal of Applied Gerontology*, *19*, 326–344.

Chyatte, S. B. (1979). *On Borrowed Time: Living with Hemodialysis*. Oradell, NJ: Medical Economics Company.

Curtin, R. B., Klag, M., Bultman, D. C., & Schatell, D. (2002). Renal rehabilitation and improved patient outcomes in Texas dialysis facilities. *American Journal of Kidney Diseases*, *40*, 331–338.

Curtin, R. B., Oberley, E., & Sacksteder, P. (1997). Compliance and rehabilitation in ESRD patients. *Seminars in Dialysis*, *10*, 52–54.

Curtin, R. B., Oberley, E., Sacksteder, P., & Friedman, A. (1996). Differences between employed and non-employed dialysis patients. *American Journal of Kidney Diseases*, *27*, 533–540.

Diaz-Buxo, J. A., Lowrie, E. G., Lew, N. L., Zhang, H., & Lazarus, J. M. (2000). Quality-of-life evaluation using short form 36: Comparison in hemodialysis and peritoneal dialysis patients. *American Journal of Kidney Diseases*, *35*, 293–300.

Drees, A., & Gallagher, E. B. (1981). Hemodialysis, rehabilitation, and psychological support. In: N. B. Levy (Ed.), *Psychonephrology I: Psychological Factors in Hemodialysis and Transplantation* (pp. 133–146). New York: Plenum.

Evans, R. W., Blagg, C. R., & Bryan, F. A., Jr. (1981). Implications for health care policy: A social and demographic profile of hemodialysis patients in the United States. *Journal of the American Medical Association*, *245*, 487–491.

Fox, E. (1997). Predominance of the curative model of medical care: A residual problem. *Journal of the American Medical Association*, *278*, 761–763.

Goldberg, A. P., Geltman, E. M., Hagberg, J. M., Gavin, J. R., Delmez, L. A., Carney, R. M., Naumowicz, A., Oldfield, M. H., & Harter, H. R. (1983). Therapeutic benefits of exercise training for hemodialysis patients. *Kidney International*, *24*(Suppl. 16), S303–S309.

Goldberg, A. P., Hagberg, J. M., Delmez, J. A., Florman, R. W., & Harter, H. R. (1979). Effects of exercise on coronary risk factors in hemodialysis patients. *Proceedings of the Clinical Dialysis and Transplantation Forum*, *9*, 39–43.

Hakim, R. M., Depner, T. A., & Parker, T. F., III (1992). Adequacy of hemodialysis. *American Journal of Kidney Diseases*, *20*, 107–123.

Hays, R. D., Kallich, J. D., Mapes, D. L., Coons, S. J., & Carter, W. B. (1994). Development of the Kidney Disease Quality of Life (KDQOL™) instrument. *Quality of Life Research*, *3*, 329–338.

Holley, J. L. (1994). Vocational rehabilitation of chronic dialysis patients: An overview of the past, present, and future. *Seminars in Dialysis*, *7*, 351–354.

Kaplan, S. L., & Schenkman, M. (Eds) (2000). Quality of life. *Neurology Report*, *24*(4), 126–158.

Kimmel, P. L., Peterson, R. A., Weihs, K. L., Simmens, S. J., Boyle, D. H., Umana, W. O., Kovac, J. A., Alleyne, S., Cruz, I., & Veis, J. H. (1996). Psychologic functioning, quality of life, and behavioral compliance in patients beginning hemodialysis. *Journal of the American Society of Nephrology*, *7*, 2152–2159.

Kobrin, S. M., Kimmel, P. L., Simmens, S. J., & Reiss, D. (1991). Behavioral and biochemical indices of compliance in hemodialysis patients. *ASA10 Transactions*, *37*, M378–M380.

Kurtin, P. S., Davies, A. R., Meyer, K. B., DeGiacomo, J. M., & Kantz, M. E. (1992). Patient-based health status measures in outpatient dialysis. *Medical Care*, *30*, MS136–MS149.

Kutner, N. G. (1987). Social worlds and identity in end-stage renal disease (ESRD). In: P. Conrad & J. A. Roth (Eds), *Research in the Sociology of Health Care* (Vol. 6, pp. 33–71). Greenwich, CT: JAI.

Kutner, N. G. (1998). Race/ethnicity, support/encouragement, and rehabilitation outcomes in incident PD patients. *Peritoneal Dialysis International*, *18*(Suppl. 1), S73.

Kutner, N. G. (2001). Improving compliance in dialysis patients: Does anything work? *Seminars in Dialysis*, *14*, 324–327.

Kutner, N. G., Brogan, D., & Fielding, B. (1991). Employment status and ability to work among working-age chronic dialysis patients. *American Journal of Nephrology*, *11*, 334–340.

Kutner, N. G., Curtin, R. B., Oberley, E., & Sacksteder, P. (1997). Fulfilling the promise: Linking rehabilitation interventions with ESRD patient outcomes. *Dialysis & Transplantation*, *26*, 282–292.

Kutner, N. G., Fielding, B., & Brogan, D. (1993). Quality of life for elderly dialysis patients: Effects of race and mode of dialysis. In: D. G. Oreopoulos, M. F. Michelis & S. Herschorn (Eds), *Nephrology and Urology for the Aged Patient* (pp. 263–276). Boston: Kluwer.

Kutner, N. G., Zhang, R., & McClellan, W. M. (2000). Patient-reported quality of life early in dialysis treatment: Effects associated with usual exercise activity. *Nephrology Nursing Journal*, *27*, 357–367.

Levin, N. W., Lazarus, J. M., & Nissenson, A. R. (1993). Maximizing patient benefits with epoetin alpha therapy: National cooperative rHu erythropoietin study in patients with chronic renal failure – an interim report. *American Journal of Kidney Diseases*, *22*, 3–12.

Lore, G. (1997). The AAKP discovers patients want a voice in their own treatment – a survey on the NKF–DOQI Guidelines. *For Patients Only*, *10*(4), 14–16.

Lundin, A. P. (1985). Expectations for long-term survival of ESRD patients. In: N. G. Kutner, D. D. Cardenas & J. D. Bower (Eds), *Rehabilitation and the Chronic Renal Disease Patient* (pp. 35–51). New York: SP Medical.

McMurray, S. D., Johnson, G., Davis, S., & McDougall, K. (2002). Diabetes education and care management significantly improve patient outcomes in the dialysis unit. *American Journal of Kidney Diseases*, *40*, 566–575.

Medical Education Institute and Contributing Editors (1994). Exemplary practices in renal rehabilitation: The winners are in! *Renal Rehabilitation Report*, *2*(4), 1–16.

Medical Education Institute and Contributing Editors (1995). Exemplary practices in renal rehabilitation: Meet the winners! *Renal Rehabilitation Report*, *3*(6), 1–16.

Meyer, K. B., Espindle, D. M., DeGiacomo, J. M., Jenuleson, C. S., Kurtin, P. S., & Davies, A. R. (1994). Monitoring dialysis patients' health status. *American Journal of Kidney Diseases*, *24*, 267–279.

Mohr, P. E., Neumann, P. J., Franco, S. J., Marainen, J., Lockridge, R., & Ting, G. (2001). The case for daily dialysis: Its impact on costs and quality of life. *American Journal of Kidney Diseases*, *37*, 777–789.

Neumann, M. E. (1997). OIG: Know your numbers brochures scarce in dialysis units. *Nephrology News & Issues*, *11*(6), 8–10.

Painter, P. (1993). Exercise expectations: Are ESRD patients receiving mixed messages? *Contemporary Dialysis & Nephrology*, *14*(5), 17, 21–22.

Painter, P., Carlson, L., Carey, S., Paul, S. M., & Myll, J. (2000). Physical functioning and health-related quality-of-life changes with exercise training in hemodialysis patients. *American Journal of Kidney Diseases*, *35*, 482–492.

Painter, P., & Johansen, K. (Eds) (1999). Physical functioning in end-stage renal disease. *Advances in Renal Replacement Therapy*, *6*(2), 107–194.

Painter, P., & Moore, G. E. (1994). The impact of recombinant human erythropoietin on exercise capacity in hemodialysis patients. *Advances in Renal Replacement Therapy*, *1*(1), 55–65.

Painter, P., & Zimmerman, S. W. (1986). Exercise in end-stage renal disease. *American Journal of Kidney Diseases*, *7*, 386–394.

Peitzman, S. J. (1996). Science, inventors, and the introduction of the artificial kidney in the United States. *Seminars in Dialysis*, *9*, 276–281.

Pellegrino, E. D., & Thomasma, D. C. (1997). *Helping and Healing*. Washington, DC: Georgetown University Press.

Porter, G. A. (1994). Rehabilitation for end-stage renal disease patients: Reflections on a new treatment paradigm. *Seminars in Dialysis*, *7*, 313–314.

Raj, D. S. C., Charra, B., Pierratos, A., & Work, J. (1999). In search of ideal hemodialysis: Is prolonged frequent dialysis the answer? *American Journal of Kidney Diseases*, *34*, 597–610.

Rasgon, S., Schwankovsky, L., James-Rogers, A., Widrow, L., Glick, J., & Butts, E. (1993). An intervention for employment maintenance among blue-collar workers with end-stage renal disease. *American Journal of Kidney Diseases*, *22*, 403–412.

Rettig, R. A. (1980). *Implementing the End-Stage Renal Disease Program of Medicare*. Santa Monica: Rand.

Rettig, R. A. (1995). Measuring functional and health status and health-related quality of life in end-stage renal disease patients: The Institute of Medicine's efforts in perspective. *Seminars in Dialysis*, *8*, 198–200.

Rocco, M. V., & Burkart, J. M. (1993). Prevalence of missed treatments and early sign-offs in hemodialysis patients. *Journal of the American Society of Nephrology*, *4*, 1178–1183.

Ronco, C., Conz, P., Bosch, J. P., Lew, S. Q., & La Greca, G. (1994). Assessment of adequacy in peritoneal dialysis. *Advances in Renal Replacement Therapy*, *1*, 15–23.

Sadler, J. (1996). Trying to measure quality? Ask the patient: IOM offshoot group promotes use of questionnaires. *Nephrology News & Issues*, *10*(10), 19 & 28.

Schrier, R. W., Burrows-Hudson, S., Diamond, L., Lundin, A. P., Michael, M., Patrick, D. L., Peters, T. G., Powe, N. R., Roberts, J. S., Sadler, J. H., Siu, A. L., Lohr, K. N., & Rettig, R. A. (1994).

Measuring, managing, and improving quality in the end-stage renal disease treatment setting: Committee statement. *American Journal of Kidney Diseases*, *24*, 383–388.

Sehgal, A. R., O'Rourke, S. G., & Snyder, C. (1997). Patient assessments of adequacy of dialysis and protein nutrition. *American Journal of Kidney Diseases*, *30*, 514–520.

Sherman, R. A. (1996). Non-compliance in dialysis patients. *Nephrology News & Issues*, *10*(6), 36–38.

Stark, Representative P. (1993). The End Stage Renal Disease Program (letter to the editor). *The New England Journal of Medicine*, *329*, 140.

Stivers, A. (1996). How an exercise program can benefit patients and the dialysis facility. *Nephrology News & Issues*, *10*(5), 39–40.

Twardowski, Z. J. (2003). We should strive for optimal hemodialysis: A criticism of the hemodialysis adequacy concept. *Hemodialysis International*, *7*(1), 5–16.

U.S. Renal Data System (USRDS) (2002). *USRDS 2002 Annual Data Report*. Bethesda, MD: National Institute of Diabetes and Digestive and Kidney Diseases, The National Institutes of Health.

Walters, B. A. J., Hays, R. D., Spritzer, K. L., Fridman, M., & Carter, W. B. (2002). Health-related quality of life, depressive symptoms, anemia, and malnutrition at hemodialysis initiation. *American Journal of Kidney Diseases*, *40*, 1185–1194.

Ware, J. E., Jr., & Sherbourne, C. D. (1992). The MOS 36-item short-form health survey (SF-36). *Medical Care*, *30*, 473–483.

DELIVERING LONG-TERM CARE IN A CHANGING ENVIRONMENT: THE IMPACT OF MANAGED CARE IN THE UNITED STATES

Teresa L. Scheid and Diane L. Zablotsky

ABSTRACT

Long-term care has increasingly been subject to mechanisms to manage care in order to control costs and meet institutional demands for cost containment and efficiency. Fundamentally, managed care seeks to limit access to services that are deemed costly, intensive and/or long-term. However, long-term care by definition requires continuity of care across diverse service sectors. In order for managed long-term care to work, these sectors must be well integrated and able to share information about client needs. In this paper we examine the growth of managed care in the long-term care sector. We have collected data from long-term care agencies in adjoining counties (one urban, the other suburban) at three points in time (1997, 1999, and 2001) and are in a position to describe changes in the types of agencies providing long-term care, and the degree of managed care penetration in our sample. We also collected data on administrators' evaluations of managed care, and their perceptions of the effect of managed care on services and service system integration. We conclude by discussing the future of long-term care.

Reorganizing Health Care Delivery Systems: Problems of Managed Care and Other Models of Health Care Delivery
Research in the Sociology of Health Care, Volume 21, 125–140

ISSN: 0275-4959/doi:10.1016/S0275-4959(03)21007-3

INTRODUCTION

The health care sector has undergone profound organizational change in response to wider social forces promoting the rationalization, corporatization, and commodification of health care (Alexander & D'Aunno, 1990; Light, 1997; Scott et al., 2000). Long-term care, including institutional and non-institutional services for a number of populations with chronic care needs, has not escaped these institutional forces promoting commodification and corporatization (Estes et al., 1993). At the same time, the aging of the population, the increasing numbers of adults with chronic care conditions, and the diversity of long-term service needs dictate growing expenditures for long-term care (Zablotsky et al., 1999). Providers must meet these growing demands for services within an institutional environment where expenditures for such care are limited by the emphasis on cost-containment. This is especially true in the United States, where health care services are subject to a variety of mechanisms to manage care. In this paper we provide a description of how long-term care services have changed in two U.S. communities: one urban and the other suburban. Longitudinal data collected at three points in time (1997, 1999, and 2001) are used to examine the changing long-term care sector (i.e. what types of organizations have survived and what new forms have emerged) and the impact of managed care on service provision and system integration.

BACKGROUND AND SIGNIFICANCE

The most noticeable effect of the rationalization of long-term care services has been privatization and the growth of large, complex corporations which manage long-term care services – a process referred to as consolidation. Privatization involves the shift to for-profit, or proprietary ownership, as well as the use of private insurance and self pay, with increased reliance upon market criteria to distribute services (Estes et al., 1993). Managed care is another consequence of the rationalization of health care. Increasingly, both private and public long-term care entities are turning to managed care to help control costs and meet institutional demands for cost containment and efficiency.

Managed care is any technique used by purchasers of health care (both private and public entities which finance care) to authorize services prior to the provision of care. The most commonly utilized mechanisms to manage care include gatekeeping, case management, utilization review, and use of case rates or capitation. Kane et al. (1998, p. 235) differentiate managed care from risk-based managed care, which occurs when a managed care organization is at financial risk to deliver services within a fixed prepaid price (i.e. capitation or case rates). It is obviously difficult to determine an appropriate amount for long-term care, which involves both acute and chronic conditions, as well as medical and social support services. Long-term care by definition requires continuity of care across diverse service sectors (included formal and informal, medical and social supports, acute care and care for more chronic conditions). In order for managed long-term care to be effective, these sectors must be well integrated and able to share information about client needs. Yet healthcare in the U.S. is not well coordinated or integrated (Anderson & Knickman, 2001).

Managed care entities authorize service on the basis of medical necessity and demonstrated efficacy; that is, reimbursements are limited to where some type of improvement can be measured. More fundamentally, managed care seeks to limit access to services that are deemed costly, intensive and long-term; social support services and rehabilitation are less likely to be authorized. Efforts to coordinate care are also not reimbursable (Anderson & Knickman, 2001). Consequently, managed care results in the medicalization of care. In addition to limiting access to services deemed costly, managed care is also characterized by an emphasis on outcomes assessment. However, it is difficult to define and measure appropriate outcomes for long-term care. It is obvious that the underlying logic of managed care is potentially incompatible with long-term care, as a result most Medicaid managed care programs have excluded long-term care, with the exception of the S/HMO and the PACE programs (Freidland & Feder, 1998).

Elsewhere we have described the critical challenges facing providers of long-term care services in the U.S.: the aging of the population, the growing number of younger adults with chronic and/or disabling conditions, the diverse service needs of long-term care recipients, and limitations on services (Zablotsky et al., 1999). In response to these challenges, the long-term care sector has moved from a reliance on institutionally-based care (primarily nursing homes) to home and community-based services (HCBS) (Estes et al., 1993). The underlying logic for HCBC is that these services are less costly (Kane et al., 1998; Szasz, 1990). However, there is little evidence that HCBC has produced cost savings (Weissert et al., 1988) although recipients are generally happier in home and community environments (Kane et al., 1998). There has also been growth in assisted living facilities, a group residential model where residents reside in apartments or complexes that offer a full range of personal care, nursing services, house keeping and congregate meals (Kane, 1995). As with home health care and other forms of home-based assistance programs, there is tremendous variation from state to state in regulations governing facilities.

There is a notable lack of any policy governing long-term care developments; this is especially true of the United States (Weiner et al., 2001). The only noticeable trend has been what Estes and Linkins (1997) describe as the Devolution

Revolution, or the movement from Federal to State control over long-term care. Devolution has occurred because most long-term care needs are paid for by Medicaid, which is controlled by states. At the same time, Medicare does pay for some home care for the elderly; consequently, home health care has been one way for states to protect scarce Medicaid dollars. In 1994, the total long-term care expenditures were over $100 billion; 44% of those costs were paid for by Medicaid funds, 16% by Medicare, and families and individuals paid for 33% of the costs (Cohen, 1998). It is no surprise that long-term care is one of the fast growing budgetary items for states, and while there has been much variability in strategies (Johnson, 1997; Weiner & Stevenson, 1998), managed care is seen as one way to make systems more efficient. There have also been federal incentives for managed long-term care; the Balanced Budget Act of 1997 established the Medicare + Choice program which was intended to encourage beneficiaries to choose a capitated health plan rather than a traditional fee for service plans (Feder & Moon, 1998).

Section 115 waivers allow states to contract with managed care organizations (Kane et al., 1998). In order to effectively manage long-term care, Medicaid and Medicare funds must be merged to allow for the full range of acute and long-term, medical and social support services to be offered to long-term clients. This kind of integration will ultimately require plans that include a broad array of services, specialization in chronic care, and sophisticated case management (Schlesinger & Mechanic, 1993) as well as some form of centralized oversight to reduce local variability (Sparer, 1996).

As described in great detail by Estes et al. (1993) the long-term care sector is highly fragmented. Prospective payment systems have resulted in quicker releases of patients, and increased reliance upon home health care. Estes et al. (1993) found that home health care agencies were more likely to be for-profit, and to select patients who can pay for their care and whose care is likely to be profitable. Home health care is provided by lower cost workers characterized as semi-skilled (home health aid) or unskilled (chore worker), and little is known about the quality of care received by home health care agencies. At the same time, "no issue is as difficult and controversial as quality of care in nursing homes" (Estes et al., 1993, p. 107; Ray, 2000). The number of nursing home beds has not kept pace with the growth in elderly needing skilled care, and there have been significant declines in length of stay with more and more residents having higher illness acuity levels: requiring more intensive forms of care and services. Difficulties acquiring and keeping qualified staff within an industry characterized by low pay and benefits, long hours, and increasingly difficult work is endemic to the long-term care sector (Hyatt, 1997). Managed care is producing greater bureaucratic control over work processes in the effort to increase productivity and control costs (Estes et al., 1993; Szasz, 1990); the lack of skill discretion and job autonomy weaken job satisfaction and may have a negative impact on the quality of care received by residents and clients (Aiken & Sloan, 1997).

As has been argued, the only way for managed care to produce better long-term care is if it can meet its promise of providing greater system coordination and efficiency (Feder & Moon, 1998). Managed care plans must deal with both acute and long-term care needs, and must integrate Medicaid and Medicare dollars. Those demonstration projects that have attempted to do this (S/HMO, PACE, On Lok) have not shown particularly promising results (Branch et al., 1995; Kane et al., 1998; Knapp & Slaton, 1997), though enrollments into these plans has been low. Outside of these special programs, managed care penetration in long-term care is influenced by the presence of statewide Medicaid managed care, resident enrollment in managed care plans (Wallace et al., 2000) and the presence of interorganizational linkages (Zinn et al., 1999). These researchers view the decision to participate in managed care as a rational response to changing environmental conditions and as a rational strategy to maintain stable resource flows. In a competitive environment, organizations with limited resources need to form interorganizational ties with one another (D'Aunno & Zuckerman, 1987; Zuckerman & D'Aunno, 1990). Another advantage of interorganization linkages is enhanced opportunities for organizational learning and innovation (Goes & Park, 1997).

Service system integration also enables clients with diverse service needs to access these services (Leutz, 1999); fragmentation occurs when there are categorical funding streams that pay for one type of service but not another – or when service agencies focus on only one type of service (Provan & Sebastian, 1998). Integration allows for the types of specialized services and coordination between providers needed to deliver long-term care. In a recent issue of *Generations* (1999, *23*, pp. 57–75) a panel of experts answered questions about the integration of acute and chronic care for the elderly. A key concern was that the integration of poor services will only produce a poor system. In addition to overcoming vested interests and categorical funding programs, integration means that diverse provider groups must agree on philosophies of care and standards for care (Cohen, 1998, p. 85). Given that managed care and prospective payment has already resulted in a medicalization of long-term care (Estes et al., 1993; Kane et al., 1998), there is concern that integration will favor acute medical services over the long-term supportive services needed by clients (Cohen, 1998).

However, we know very little about the degree of integration that has occurred in the long-term care sector, or what effect integration has had on services. In their survey of managers of multilevel long-term care facilities Wallace et al. (2000) found that while administrators were concerned about quality of care, this concern was not in fact associated with managed care involvement. In this paper, we also use data collected from long-term care administrators to address the following questions:

(1) What organizational forms have survived in the long-term care sector, and what new organization forms have emerged in response to changing institutional demands?
(2) To what degree has managed care penetrated the long-term care sector? Are agencies with managed care involvement different from those without such involvement?
(3) What are administrators' attitudes toward managed care? Has it produced a better system of care, has it controlled costs, has it resulted in improved access, quality or client care?
(4) What changes have occurred in service system coordination or integration?

METHODOLOGY

Study Design

We collected data for this research in one urban and adjacent suburban county in North Carolina at three points in time: 1997, 1999, 2001. Both counties have experienced rapid population growth since the 1990s (i.e. the suburban county experienced a 60% population increase during the 1990s). We used an open ended and fixed response questionnaire to obtain information from administrators of long-term care services on the following general areas: characteristics of the agencies, characteristics of the client population, service provision and coordination, access and the social environment, and the impact of managed care. In addition to requesting actual information about the activities of the facility or agency we asked the administrators to describe elements of their service environment and their perceptions of service delivery and managed care.

We gathered information from administrators providing the following services: adult day care, home health care, residential care (including assisted living), and nursing homes. When the original sampling list was constructed in 1996, we used the following criteria to determine if a provider was a long-term care facility, organization, or agency: (a) residents served by the facility require assistance over long periods of time; or (b) residents served by the facility require assistance with an Activities of Daily Living; or (c) residents are entitled to have care paid for through Medicare or Medicaid.

We started with the telephone book and lists provided by local agencies, and then used snowball techniques to ensure that we had a comprehensive list of long-term care providers. In Table 1, we use the population frame for our study to examine changes in the types of agencies providing services in the long-term care sector. The most obvious change is that in 1999, we added assisting living

Table 1. Population Ecology Analysis of the Long-Term Care Sector (n).

	1997	Surviving	1999		Surviving	2001	
			Births	Deaths		Births	Deaths
Nursing homes							
Urban	21	21	8	0	25	6	4[a]
Suburban	15	15	0	0	13	2	2
Assisted living							
Urban	0	0	25	0	21	19	4
Suburban	0	0	8	0	8	0	0
Home health							
Urban	31	23	9	8	6	0	17
Suburban	4	4	2	0	5	1	1
Adult day care							
Urban	6	6	3	0	5	2[b]	4
Suburban	3	3	1	0	2	1	2
Totals	80	72	56	8	85	31	34

[a] Two of these nursing homes did not "die"; they evolved into assisted living facilities.
[b] One of these new adult day care centers was previously a home health agency.

facilities, responding to the relatively new growth in this arena, as well as changing state regulations for residential living facilities. In 1999, thirty-three assisting living agencies (or home care facilities which provide some kind of residential living with assistance to residents) were added to the population, and in 2001, nineteen new assisted living facilities emerged. Furthermore, these facilities demonstrate some stability; only four "deaths" occurred in 2001, whereas overall thirty-four long-term care agencies "died" in this time period.

The second most obvious finding is the turbulence in the home health care arena: we see many new home health care agencies, and a high number of organizational "deaths" in both 1999 and 2001. There is also a relatively high degree of turbulence in the adult day care arena, with a number of "births" and "deaths" in our 2001 population list. There was some degree of organizational evolution – with two nursing homes evolving into assisted living facilities and one home health agency becoming an adult day care center. Also, while not reflected in Table 1, a number of agencies with a religious affiliation "died" and no new long-term care services provided by religious organizations emerged. Another sign of turbulence in the long-term care sector is the extremely high turnover in the agencies. We contacted each agency director before sending the survey, but found that in the majority of our sample, the director was no longer there when we attempted a follow-up survey.

Sample Characteristics

Despite the original phone call to inform administrators about our study, repeated mailings and reminders, and innumerable telephone calls, we were not able to obtain a very high response rate. In 1997, we mailed questionnaires to 80 agencies and 34 were returned (42.5%); in 1999, questionnaires were mailed to 120 agencies and 39 returned (32.5%), and in 2001, 82 questionnaires were mailed and only twenty were returned (24.4%). This is partly due to turbulence in the field, and the fact that administrators in human service agencies are overwhelmed with requests for information. Attempts to complete the surveys by telephone or in person have had little impact as administrators claimed they could not spare the 1/2 hour to complete the interview. Only 15 agencies completed questionnaires at two points in time, consequently our analysis is confined to a discussion of aggregate changes in the organizational field.

When comparing the respondents (Table 2) we see that the number of home health agencies responding to the survey decreased over time and that we also had

Table 2. Characteristics of Respondents (Comparisons Over Time).

	1997 ($n = 34$)	1999 ($n = 39$)	2001 ($n = 20$)
Type of facility			
Nursing home	41%	38%	35%
Assisted living	0	26	30
Home health	47	26	10
Adult day care	12	10	25
Ownership			
Public/non profit	39%	43%	35%
Profit	61	57	65
Funding source			
Public (federal/state/local)	53%	44%	43%
Private (private insurance/self pay)	47	56	57
Change in client census			
Increase	56%	44%	55%
Decrease	15	17	5
Remain the same	29	39	40
Mean % clients over 85			
Nursing homes	–	25.6	31.5
Assisted living	–	32.8	37.3
Home health care	–	23.0	19.5
Adult day care	–	9.5	24.2

Table 2. *(Continued)*

	1997 ($n = 34$)	1999 ($n = 39$)	2001 ($n = 20$)
Mean % clients black ($p = 0.010$)			
Nursing homes	17.7	33.6	18.2
Assisted living	–	19.6	12.2
Home health care	29.7	39.4	19.0
Adult day care	48.5	44.0	57.8
Mean % clients female ($p = 0.063$)			
Nursing homes	73.9	74.2	69.5
Assisted living	–	64.6	73.3
Home health care	52.7	67.4	59.0
Adult day care	62.7	72.0	65.8
Mean No. full time providers ($p = 0.007$)			
Nursing homes	65.2	59.9	66.9
Assisted living	–	15.1	24.0
Home health care	34.1	28.7	62.5
Adult day care	61.5	7.0	3.0
Overall staff/client ratio			
Nursing homes	8.4	8.9	8.8
Assisted living	–	5.6	11.0
Home health care	7.9	9.9	65.0[a]
Adult day care	6.5	5.0	5.2
Averaged number of shortages (adding serious shortages in money, staff, space, equipment, and other – summary scale from 0 to 2)			
Nursing homes	1.9	1.4	1.4
Assisted living	–	1.3	0.83
Home health care	1.0	1.1	0.50
Adult day care	1.75	2.0	2.2

[a] Only one home health care agency provided information on staff in 2001.

a lower response rate in 2001 for the nursing homes. The samples are similar in ownership and funding source, with slight increase in the proportion of facilities which are for profit. The majority of providers in our sample have experienced an increase in their client census. In looking at client demographics and staffing patterns, we compare agencies across time. In general, agencies provided services to between 20 and 25% of residents who were over 85, and consequently need more care. Not surprisingly, adult day care agencies served the fewest number of those over 85. The majority of the residents were female (63–70%), with smaller proportions of women served by home health care ($p = 0.063$). There were also significantly more African Americans in adult day care ($p = 0.01$) with far fewer

African Americans in assisted living facilities. There is a great deal of variability in the numbers of full time providers across time and between types of agencies, but overall there is a decline from 50 full time providers to 38 between 1997 and 2001. Staff ratios also increased (overall from 7.9 in 1997 to 11.67 in 2001). Adult day care agencies faced the greatest number of serious shortages, and home health care the fewest average number of serious shortages.

MANAGED CARE IN THE LONG-TERM CARE SECTOR

In 1996, when we first developed our questionnaire, managed care was just a blip on the radar screen, and managed care was not yet a feature of the long-term care sector. Consequently, we did not ask about managed care until 1999. At this time slightly over half of the agencies who responded had some mechanism to manage care, and slightly over 60% had some mechanism to manage care in 2001 (Table 3). Agencies use a variety of mechanisms to manage care, including outcomes assessment, case management, utilization review, concurrent review and capitation (Table 3). Assisted Living facilities were least likely to have managed care, which could be due to the newness of this organizational form.

Table 3. Managed Care Penetration (%).

	1999 ($n = 38$)	2001 ($n = 20$)
Managed care penetration		
Some mechanism to manage care	56.3	61.5
No mechanism to manage care	43.8	38.5
Mechanisms to manage care		
Outcomes assessment	41.0	53.8
Case management	50.0	35.7
Utilization review	36.7	50.0
Concurrent review	26.7	30.8
Capitation	31.0	33.3
Type of facility with managed care		
Nursing home	63.6	100
Assisted living	28.6	0
Home health	70.0	100
Adult day care	50	66.7
	% Yes	% Yes
Has managed care had an effect on		
The types of clients/residents you serve?	25.0	33.3
The types of services you provide?	9.4	15.4

In 2001, the majority of agencies with managed care had specific staff positions to authorize services (data not shown). The majority of LTC providers reported that managed care had not had any effect on the types of clients or residents they serve or on the types of services they provide, although the percentage of administrators responding yes to this question increased over time. At the same time, attitudes toward managed care "softened" between 1999 and 2001; in 1999, the majority of administrators did not feel managed care had produced a better system of care, improved quality of care, or improved client outcomes while in 2001 only 20–25% of administrators had similar beliefs about managed care. At the same time, most felt managed care resulted in under service to clients, and did not feel it had produced greater control over costs. In 2001, administrators were asked if managed care had resulted in greater referrals to other agencies; only 35% disagreed, indicating that managed care had resulted in more referrals (data not shown). This may indicate greater system integration, or else cost shifting.

We next turn to a comparison of those facilities with managed care penetration, and those without managed care penetration (Table 4). Facilities with managed care had a smaller proportion of clients over 85 in both 1999 and 2001, and they served more African American clients than facilities without managed care (these

Table 4. Comparison of Facilities with Managed Care Penetration 1999 and 2001.

Managed Care	1999		2001	
	Yes ($n = 18$)	No ($n = 14$)	Yes ($n = 8$)	No ($n = 5$)
Mean No. clients over 85 (S.D.)	14.24	27.50	23.83	30.20
	(12.0)	(26.2)	(10.8)	(29.7)
Mean % black residents (s.d) ($p = 0.104$)	43.24	28.92	42.37	20.00
	(29.8)	(38.6)	(33.2)	(17.0)
Mean % female residents (S.D.)	64.00	74.15	69.53	65.80
	(17.7)	(28.4)	(12.7)	(26.4)
Mean full time providers (S.D.) ($p = 0.014$)	51.65	12.57	42.00	21.00
	(59.9)	(8.80)	(46.3)	(19.6)
Mean staff/client ratio (S.D.)	9.67	6.64	16.57	11.80
	(7.3)	(6.1)	(21.5)	(7.5)
% For profit status	44.4	42.8	75.0	60.0
% Changed ownership[a]	22.2	0	12.5	0
% Experienced funding losses	80.0	20.0	87.5	12.5
% Felt access to be inadequate	62.5	37.5	60.0	40.0

[a] All of the facilities that changed ownership had mechanisms to manage care. None of the facilities without managed care changed ownership.

Table 5. Changes in Service System Coordination.

	1997	1999	2001
How many agencies do you collaborate with?			
Mean (S.D.)	11.17	3.90	4.45
	(20.9)	(4.75)	(4.45)
How many agencies do you provide services for?			
Mean (S.D.)	24.21	15.92	23.74
	(42.1)	(36.4)	(41.5)
What % of your services do you contract out?			
Mean (S.D.)	30.62	36.08	42.30
	(41.9)	(42.8)	(47.5)

two demographic facts are related in that African Americans have lower mortality rates than Whites). Managed care facilities also have more full time providers, although the staff client ratio is higher than facilities without managed care indicating agencies with managed care serve more clients. While in 1999 slightly less than half of the managed care facilities were for profit, by 2001 75% were for profit, although 60% of the non-managed facilities were also for-profit. This is a reflection of larger forces promoting corporatization. All of the facilities that changed ownership had managed care. Agencies with managed care were much more likely to report experiencing funding losses, and were also more likely to feel access was inadequate (these perceptions did not significantly change over time).

Table 5 examines aggregate changes in service system coordination, and we find that in 1999 and 2001 agencies are less likely to collaborate with other agencies, and more likely to contract out. In 2001, we asked administrators to list those agencies with whom they maintained some kind of contractual relationship, providing six lines for them to fill in. Only six (out of 20) indicated they did not have any agencies with whom they contracted – four mentioned hospitals or doctors, five listed other long-term care agencies that provided a different service, and the majority included a variety of other social service agencies in the area (such as Department of Social Services).

CONCLUSIONS

We found a great deal of turbulence in the long-term care sector, with an increasing number of organizational births and deaths in the six years we have been tracking population changes in the long-term care sector. According to the theory of population ecology, older organizations are most likely to survive, new organizational forms are the most unstable, and more likely to die, as are organizations attempting reorganization (Baum, 1996; Peli et al., 1994). Those organizations with the highest survival rates are those which are older and have established normative legitimacy as well as high levels of external accountability. These organizations are not only most likely to persist and survive over time; they are also subject to inertia and resist change. This has certainly been true of nursing homes, and our data support these generalizations. However, Baum (1996) argues that with increased market concentration large organizations which perform a variety of roles or tasks (such as nursing homes) are more likely to fail if new, smaller, more specialist organizations are able to exploit available resources. This explains why assisted living agencies have not failed, despite their newness to the long-term care sector. Assisted living homes present an alternative to nursing home use and are more cost efficient. Home health care agencies experienced the highest rates of organizational death. This is due to their small size, and a lack of reliability and accountability, as well as a lack of established links with other agencies. It is very difficult to find data on home health care, and the industry has been subject to a wide degree of uncertainty as regulations over home health care have changed over time. In terms of the future of the long-term care sector, it would seem that the majority of existing nursing homes will certainly persist, though they are likely to modify their structures to deal with increasing numbers of frail elderly. Assisted living will become the dominant form of residential care as the number of baby boomers age, and they exercise their preference for as normal a life as possible in the face of age related disability (Kane & Kane, 2001). If home health agencies are to survive, they need to establish more links with other long-term care agencies as well as other social service agencies, and seek to increase their legitimacy vis-à-vis the other types of agencies providing long-term care.

In terms of managed care, we found that there was a substantial degree of managed care penetration with the majority of agencies reporting some mechanism to manage care. While managed care has not improved long-term care, and there is evidence of increased funding shortages under managed care, it has not had a direct effect on services or the clients served (with the interesting exception that agencies with more managed care serve more African Americans – which could be an artifact of our sample). Managed care organizations are more likely to refer out services, consistent with attempts to contain costs by shifting clients to other agencies. While our data are limited to one geographic region, and the relatively small number of agencies responding, there is clearly some cause for concern that managed care is not going to lead to an improved system of long-term care.

The conflict between long-term care and managed care is most easily articulated within the rhetoric of traditional health care policy debates over the appropriate balance between cost-containment, access, and quality of care. Costs can be controlled only at the risk of limiting access to high quality care. Since both nursing homes and home health care agencies have been widely critiqued for poor quality care, the growth of managed care strategies is a major source of concern. As described by Schlesinger and Mechanic (1993), cost sharing is least effective and equitable when applied to long-term care populations. Furthermore, there is not much room in the long-term care arena for cost savings (Romano, 1999). The only way that managed care can lead to enhanced access and/or quality is by providing for greater system wide coordination, and consequently greater efficiency.

However, our data demonstrate the forces promoting consolidation and privatization may work against the ability of managed care arrangements to promote system integration. Large corporate facilities are more able to provide a diverse array of services within a facility and often include both home health care and various forms of residential care under one organizational umbrella. However, we do not know whether in fact residents receive this full array of services, or how utilization review and capitation may work to limit the care provided to long-term care clients. Our data is obviously exploratory, but it does point to the need for more focused study of the effect of various practices to manage care on access and quality of long-term care services. A recent review of long-term care in the U.S. points to the many gaps in service provision, and argues we need policies to address both system integration and quality care (Feder et al., 2000). This is true of other countries as well, and systems with socialized healthcare are also grappling with how to provide quality long-term care in an institutional context which mandates cost containment. As put simply by Weiner et al. (2001, p. 241) "long-term care policy is largely about money – that is, how much are we willing to spend?" Quality long-term care should slow the rate of physical, emotional and social decline (Kane & Kane, 2001); in order to do this acute and long-term care, formal and informal care, medical and social support services must all be integrated. While managed care in the U.S. has not been a disaster for the long-term care sector, it is not likely to produce the kind of system we would all want to grow old in.

ACKNOWLEDGMENTS

This research was supported by a UNC-C Faculty Research Grant. Earlier versions of this paper were presented at the Southern Gerontological Society Meetings (April 2000) and the Gerontological Society of America Meetings (November 2000). We are grateful to Audrey Crowder, Ashley Dunn, and Melody Ruth for their research assistance.

REFERENCES

Aiken, L. H., & Sloan, D. M. (1997). Effects of specialization and client differentiation on the status of nurses: The case of AIDS. *Journal of Health and Social Behavior*, *38*, 203–222.

Alexander, J. A., & D'Aunno, T. A. (1990). Transformation of institutional environments: Perspectives on the corporatization of U.S. health care. In: S. S. Mick (Ed.), *Innovations in Health Care Delivery* (pp. 53–85). San Francisco, CA: Jossey-Bass.

Anderson, G., & Knickman, J. R. (2001). Changing the chronic care system to meet people's needs. *Health Affairs*, *20*, 146–160.

Baum, J. A. (1996). Organizational ecology. In: S. R. Clegg, C. Hardy & W. R. Nard (Eds), *Handbook of Organization Studies* (pp. 77–114). London: Sage.

Branch, L. G., Coulan, R. F., & Zimmerman, Y. A. (1995). The pace evaluation: Initial findings. *The Gerontologist*, *35*, 349–359.

Cohen, M. A. (1998). Emerging trends in the finance and delivery of long-term care: Public and private opportunities and challenges. *The Gerontologist*, *38*, 80–89.

D'Aunno, T. A., & Zuckerman, H. S. (1987). The emergence of hospital federations: An integration of perspectives from organization theory. *Medical Care Review*, *44*, 323–342.

Estes, C. L., & Linkins, K. W. (1997). Devolution and aging policy: Racing to the bottom for long term care. *International Journal of Health Services*, *27*, 427–442.

Estes, C. L., Swan, J. H., & Associates (1993). *The long term care crisis: Elders trapped in the no-care zone*. Newbury Park: Sage.

Feder, J., Komisar, H. L., & Niefeld, M. (2000). Long-term care in the United States: An overview. *Health Affairs*, *19*, 40–55.

Feder, J., & Moon, M. (1998). Managed care for the elderly: A threat or promise? *Generations*, *22*, 6–10.

Freidland, R. B., & Feder, J. (1998). Managed care for elderly people with disabilities and chronic conditions. *Generations*, *22*, 51–57.

Goes, J. B., & Park, S. H. (1997). Interorganizational links and innovation: The case of hospital services. *Academy of Management Journal*, *40*, 673–696.

Hyatt, L. (1997). Staffing: A multi-faceted challenge. *Nursing Homes Long Term Care Management*, *46*, 11–13.

Johnson, J. R. (1997). State and local approaches to long term care. *Annual Review of Gerontology and Geriatrics*, *17*, 112–139.

Kane, R. A. (1995). Expanding the home care concept: Blurring distinctions among home care, institutional care, and other long term care services. *The Milibank Quarterly*, *73*, 161–186.

Kane, R. A., & Kane, R. L. (2001). What older people want from longterm care, and how they can get it. *Health Affairs*, *20*, 114–127.

Kane, R. A., Kane, R. L., & Ladd, R. C. (1998). *The heart of long-term care*. New York: Oxford University Press.

Knapp, K. R., & Slaton, R. A. (1997). Managed care: A call to action. *Nursing Homes Long Term Care Management*, *46*, 38–44.

Leutz, W. N. (1999). Five laws for integrating medical and social services: Lessons from the U.S. and the U.K. *The Milibank Quarterly*, *77*, 77–110.

Light, D. W. (1997). The rhetorics and realities of community health care: The limits of countervailing powers to meet the health care needs of the twenty-first century. *Journal of Health Politics Policy and Law*, *22*, 106–145.

Peli, G., Bruggeman, J., Masuch, M., & O'Nuallian, B. (1994). A logical approach to formalizing organizational inertia. *American Sociological Review*, *59*, 571–593.

Provan, K. G., & Sebastian, J. G. (1998). Networks within networks: Service link overlap, organizational cliques, and network effectiveness. *The Academy of Management Journal*, *41*, 453–483.

Ray, W. A. (2000). Improving quality of long-term care. *Medical Care*, *38*, 1151–1153.

Romano, M. (1999). Pennies from heaven. *Contemporary Longterm care*, *18*, 46–53.

Schlesinger, M., & Mechanic, D. (1993). Challenges for managed competition from chronic illness. *Health Affairs*, *12*, 123–137.

Scott, R. W., Ruef, M., Mendel, P., & Caroneer, C. A. (2000). *Institutional change and organizational transformation of the healthcare field*. Chicago, IL: University of Chicago Press.

Sparer, M. S. (1996). Medicaid managed care and the health reform debate: Lessons from New York and California. *Journal of Health Politics, Policy and Law*, *21*, 433–460.

Szasz, A. (1990). The labor impacts of policy change in health care: How federal policy transformed home health organizations and their labor practices. *Journal of Health Policy, Politics and Law*, *115*, 191–219.

Wallace, S., Cohn, J., Schnelle, J., Kane, R., & Ouslander, J. G. (2000). Managed care and multilevel long-term care providers: Reluctant partners. *The Gerontologist*, *40*, 197–205.

Weiner, J. A., Estes, C. L., Goldson, S. M., & Goldbert, S. (2001). What happened to long-term care in the health reform debate of 1993–1994? Lessons for the future. *The Milibank Quarterly*, *79*, 207–252.

Weiner, J. M., & Stevenson, D. G. (1998). State policy on long-term care for the elderly. *Health Affairs*, *17*, 81–100.

Weissert, W., Cready, C. M., & Pawelack, J. E. (1988). The past and future of home and community based long-term care. *The Milibank Quarterly*, *66*, 309–388.

Zablotsky, D. L., Scheid, T. L., & Roberson, T. (1999). Preparing for an uncertain future: Assessing the service environment and the challenges of providing long-term care. *Research in the Sociology of Health Care*, *16*, 49–169.

Zinn, J. S., Mor, V., Castle, N., Intrator, O., & Brannon, D. (1999). Organizational and environmental factors associated with nursing home participation in managed care. *Health Services Research*, *33*, 1753–1767.

Zuckerman, H. S., & D'Aunno, T. A. (1990). Hospital alliances: Co-operative strategy in a competitive environment. *Health Care Management Review*, *15*, 21–30.

PART III: LESSONS FROM OTHER COUNTRIES

MAKING REFORM LOCALLY: GENERAL PRACTITIONERS, HEALTH CARE MANAGERS AND THE "NEW" BRITISH NATIONAL HEALTH SERVICE

Katherine Clegg Smith

ABSTRACT

The National Health Service is key to Britain's welfare state, and has been subject to repeated reform initiatives. Such reforms rarely "fix" the problems for which they are introduced, but evaluations have neglected the significance of local action. Reform implementation involves local translation of politically contextualized ideas into workable practice. I focus on implementation processes and the role of professions. Ethnographic data reveal local actors engaging with policy objectives to protect existing structures within the boundaries of official reform rhetoric. Actors employ multiple strategies to maintain existing systems. Rather than "failing," policy is made through localized collaboration.

INTRODUCTION

When an observer looks behind the façade of formal institutions to the way they are realized in practice, he or she gains a considerably more complex and precise view of the exercise of

Reorganizing Health Care Delivery Systems: Problems of Managed Care and Other Models of Health Care Delivery
Research in the Sociology of Health Care, Volume 21, 143–162

ISSN: 0275-4959/doi:10.1016/S0275-4959(03)21008-5

power or influence than is provided by official charters, tables of organization and legislation (Freidson, 1993, p. 56).

The New NHS: Modern, Dependable

In April 1997, Britain elected a new Labour (left wing) government under Prime Minister Tony Blair. The landslide election signified the end of 18 years of Conservative (right wing) domination, with a Labour manifesto of major reform initiatives to bolster the welfare state. At the heart of Labour's promises for a "new" Britain was a move away from the largely unpopular Conservative government National Health Service (NHS) reforms of the 1980s and 1990s in which market-based reforms were superimposed onto the existing collectivist system (Clarke, Cochrane & McLaughlin, 1994).

The key Conservative initiative for the purposes of the present discussion was the introduction of market-based structures for general practitioners (GPs) based loosely on U.S. managed care principles. An "internal market" between purchasers and providers of care was established in which GPs were encouraged to become "fundholders." Such "fundholding" GPs were allocated resources with which to purchase a range of services for their patients. The incentive was that any savings made on such purchases could be applied to other areas of the practice, and the system was designed to encourage fundholding GPs to be more critical of the cost of care and more frugal than under the existing, demand-driven system. The internal market was also designed to foster a philosophy of competition between local doctors, as well as to enhance professional involvement in resource management. The effect of the reforms was such that by the late 1980s and early 1990s, a "managerial" or "business" ideology had begun to permeate the notions of British primary care (Clarke, Cochrane & McLaughlin, 1994).

Soon after the 1997 election, the Labour government released its plans for a radical reorganization of the NHS based upon the ideology of professional collaboration rather than free market competition under the title, *The New NHS: Modern, Dependable* (Department of Health, 1997). The policy did not propose to re-instigate pre-fundholding institutions, but rather introduced new organizational structures known as Primary Care Groups (PCGs) that were then charged with the work of resource allocation for their community. PCGs would be led by local professionals (primarily GPs) and other "stakeholders" (such as nurses, social services representatives and voluntary sector representatives) working in and for a local population of approximately 100,000 patients. The preparation process for PCGs mirrors the four key strategies for regulating clinical activity within

U.S. managed care organisations: financial incentives, medical practice profiling, disease management and utilisation review (Robinson & Steiner, 1998).

Reform in the NHS

The history of the NHS reveals a system that has been undergoing change initiatives constantly since its initial formation. Most modern governments have experienced a "recent history of exceptional growth and change" (March & Olsen, 1983, p. 281), and much of the work of contemporary government involves managing such change. With advances in medical technology and improvements in public health, so public expectations regarding health and health care continue to expand. The central position adopted by the British state with regards to resourcing and providing health care has meant that, alongside increasing public expectations, have come increasing demands on the state. These increasing demands have led to numerous reconfigurations of the system. It has, however, been argued that much of the administrative reform that has been introduced within the NHS reflects rhetorical activity, rather than any actual changes to practice (March & Olsen, 1983). To this extent, the concept of "reform" has become increasingly problematized within the sociological literature, and subject to more detailed analysis (Ferlie, 1997).

Freidson (1975), perhaps somewhat cynically, observed that every period in history tends to find it necessary to declare itself in a state of crisis, and the increasing normalization of change seems likely to intensify this perceived need. The experienced need for change provides policy makers with an opportunity to present reform initiatives as "solutions" to existing contradictions or difficulties. Health care reform may be used demonstrate legitimacy, rather than to achieve any "real' change." Thus, outside of any evaluation of their effectiveness, reform measures can be politically expedient through their very existence since reforms offer the promise of improvement.

British policy makers have often attempted to lend authority to reform proposals by incorporating professionals and thus benefitting from high levels of trust on the part of the public. The incorporation of professionals into the governance of health care may serve to temper public resistance, and to assist in devolving responsibility for unpopular action away from central government (Ham, 1998). Such attempts at professional engagement can reveal a source of potential conflict and contradiction to the extent that reform initiatives can be conceived as assertions of greater state control over clinical practice (Hughes & Allen, 1993).

The translation of abstract policy ideas into workable practices can reveal much about the groups charged with reform implementation. Relatively few studies have, however, given detailed consideration to the micro-level interaction that is key to policy implementation (Griffiths & Hughes, 1999). An informed understanding of policy reform should incorporate the significance of the process of organizational change in and of itself, as well as the idea of reform as a response to either changing conditions or a shifting context. The role of local actors as "lenses" for initiatives may result in unintended and unforeseen consequences, the effects of which often seem to be dismissed as policy "shortcomings" or "failures." Quite different perspectives may be gleaned from considerations of the mechanisms established to deliver the initiatives, as compared to those outlined in official policy documentation.

I spent approximately eighteen months in one English health authority[1] (given the pseudonym "Casterdale") following the implementation of policy initiatives at the local level. The data that I draw upon comes from participant observation of reform implementation meetings between doctors and managers, and ethnographic interviews with participants in the change measures. This study focuses on local negotiations, and is concerned with the way that policy rhetoric is implemented locally, particularly the role played by the GP profession. The data reveal that existing local structures of professional involvement were shown to shape the interpretation and adoption of the decentralized structures as officially outlined by policy makers.

My research provided a window into the micro-processes of policy implementation and organizational change. The data indicate that local GPs and managers often manipulated the policy agenda to limit any actual changes to working practice and to protect existing local networks and structures. Local implementation was largely defined by active efforts on the part of both managers and professionals to maintain co-operation and consensus, and to strengthen existing relationships. I suggest that there are many reasons why one might have anticipated the implementation of this reform initiative to be characterized by conflict between GPs and local NHS managers. However, contrary to expectations, the data reveal a complex portrait of co-operation, consensus building, and only occasional conflict both between and within the different groups.

THE WORK OF TRANSLATING POLICY RHETORIC INTO WORKABLE STRUCTURES

Reform proposals such as *The New NHS* exist initially in the form of political ideas and rhetoric. It is therefore always necessary that considerable work be undertaken to translate such ideas into the practical structures through which the objectives of the changes can be realized. Rose and Miller (1992) coined the term "governmental technologies" to describe the mechanisms by which centrally initiated rules are operationalized. One such technology is to engage local professionals in the implementation of policy. Reform initiatives are, however, characteristically conceptualized as undermining professional claims, rather than as offering opportunities for development. Hughes (1964) proposed that although professionals have tended to benefit from changes to organization of their work, they have been made uneasy by such transformations.

The role of GPs in the implementation of NHS reform is of particular interest since they are central to the system's effectiveness through their role as "gatekeepers" (or filters) to more expensive, secondary services (Light, 1993). GPs' position on "the front-line" with patients makes them potentially influential players in determining how well the system operates. General practitioners are increasingly being charged with the practical implementation of reform measures, and are both formally and informally called upon to explain and interpret the likely effects of any changes for their community of patients.

GPs were therefore placed at the heart of the 1997 reforms, *The New NHS: Modern, Dependable*. The changes introduced to the organization of general practice under the Conservative system had not been generally popular with either the public or many doctors. Thus, *The New NHS* abandoned the "market-based" fundholding measures, while continuing to expand the role of GPs in determining resource allocation.

The history of the NHS has often been described as one of near-constant conflict between the policy makers and the medical profession, and considerations of health care systems have frequently emphasized the divisions between managers and health care professionals when accounting for the limited success of reform initiatives (Ham, 1992). Managers and professionals are frequently conceived as being at odds with one another because of assumptions of fundamentally different underlying philosophies of the two groups (Coombs, 1987), objectives (Drummond, 1998) and operational agendas, with conflicting rather than complementary roles. The likelihood of professional/managerial conflict is also conceptualized as intensifying during periods of organizational change (Flynn, 1992) – such as that brought about by *The New NHS*. Organizational change contributes to great uncertainty, and thus poses opportunities for existing relationships between different groups to be challenged.

The professional history of general practice has been shaped by the relatively low status of GPs in relation to hospital doctors (Honigsbaum, 1979), and GPs' historic financial dependence on their patients (Calnan & Gabe, 1991). Thus, unsurprisingly, there was initially a significant amount of GP opposition to the 1997 reform proposals. Over time, however, this resistance largely dissipated such that by the time it came to implement the measures in

Casterdale, all of the local GPs were essentially in compliance with the reform measures.

Thus, in relation to *The New NHS*, GPs seemed to have quite quickly appreciated opportunity to influence reform measures as an alternative to witnessing the processes of "deprofessionalisation" (Haug, 1973), or even "proletarianisation" (McKinlay & Arches, 1985) that are often predicted within sociological accounts. It may be that rather than having to "drag" professionals into co-operation with reform initiatives, these GPs joined willingly when they recognized that such involvement was a potential means by which to strengthen their professional claim.

Data from this study indicate that the perception of underlying differences between managers and GPs has permeated the expectations of these local actors. In one interview, a manager described efforts on the part of both managers and GPs to establish a basis of understanding and co-operation. He constructed an account of the two sectors (managers and GPs) inhabiting completely different worlds, but proffered that managers and GPs were seeking to understand one another's position and to work together.

> Health Authority Manager – A local doctor and I gave a joint presentation (in relation to the future of PCGs). He imagined what the big organization (the health authority) was like and he presented the health authority perspective. I presented the GP perspective – the big structured organization used to working in an accountability structure, used to handling decision-making processes and so on, comes together with the small, autocratic, the doctor being dictator in his own realm – and these two come together.
>
> (*Interview data*)

To the extent that managers have only limited autonomy in policy implementation (Griffiths & Hughes, 1999), so it may serve them to appear as pliable as possible in the process of local application. Indeed, managers might secure legitimacy by emphasizing the extent to which their role is to simply apply the rules as determined by the government rather than working for any strong preservation of particular measures. In this way, managers are able to protect their local position: defending against charges of lay intrusion into the care domain on the part of medical professionals, as well as to the strength of public opinion over matters relating to health and health care.

Contrary to existing theory that professionals interpret managerial involvement as problematic (Flynn, 1992; Greenfield & Nayak, 1996), it would seem that certain GPs also appreciated the potential benefits from acknowledging and appropriating a managerial perspective within this change process. FitzGerald (1996) proposes that "clinical management" roles that have traditionally been seen as outside the realm of "real" care, are increasingly being used to augment doctors' involvement in informing major decisions that shape the delivery of health care. Thus, reform measures that blur the boundaries between managerialism and professionalism may provide scope for GPs to be increasingly influential in decision making (Griffiths & Hughes, 1999), rather than challenging the core of professional power.

The following sections explore various aspects of the reform implementation between managers and GPs in pursuit of these perspectives.

Local Implementation of Reform: GPs and Managers Working to Create and Maintain Consensus

The centrality of GPs to the 1997 reform initiatives is established within the policy rhetoric.

> For the first time in the history of the NHS all the primary care professionals, who do the majority of prescribing, treating and referring, will have control over how resources are best used to benefit patients (The New NHS Department of Health, 1997, p. 37).

The reform's stated aims of augmented professional participation alongside strengthened systems of clinical governance and decentralized decision making depend on enlisting professional support for and engagement with the measures. At the same time, local health care managers also have a key role to play in the translation of policy objectives into working structures.

The sociological literature predicts conflict between managers and GPs over reform initiatives (Coombs, 1987; Dopson, 1996; Flynn, 1992; Miller, 1970) and indeed, the perceived failure of many reform initiatives is sometimes attributed to such animosity. The data in this study, however, strongly suggest that rather than openly opposing one another, managers and GPs in Casterdale engaged various strategies of accommodation and consensus maintenance during the period of reform implementation. Observational data from the meetings revealed considerable work on the part of both GPs and Health Authority managers to protect consensual forms of engagement and to avoid encounters with potentially problematic issues.

This is not to suggest, however, that managers and GPs were necessarily collaborating towards commonly-held goals. The interaction observed instead suggests that the two groups of actors developed techniques through which they could independently pursue their own agendas while avoiding facing overt confrontation. Thus, the dualist approach that sets up professionalism and bureaucracy as necessarily in conflict may be over simplistic (Larson, 1977, p. 191). These data suggest that bureaucracy (in the form of managers) and professionalism (in the form of the GPs) are sometimes interdependent in terms of attaining and maintaining social control.

Consensus Strategies: "Fudging" Definitions and Talking Past Each Other

One might assume that during reform implementation it would be necessary for local actors to collectively establish an understanding of the reform objectives, as well as to determine a set of common aims and rules of operation. Indeed the existence of distinct cultures in different parts of an organization can impede organizational communication (Deal & Kennedy, 1988). Establishing "common-ground" between key actors is therefore usually understood as essential for successful collaboration. The existence of official policy goals and targets allow members of an organization to display adherence to collective norms.

Thus, one might expect managers and GPs (who had been charged with implementing change and forming new organizational structures) to work to establish a set of norms and common expectations. In Casterdale, however, there was little evidence of this. On the contrary, it often appeared that managers and GPs were talking different languages in relation to the reforms, without the actors seemingly reacting negatively to this "Tower of Babel" scenario. Actors worked to avoid identifying a detailed, collective target or agenda at the local level, and instead made numerous references to the policy rhetoric. When interviewed, one of the higher level managers commented on the lack of a precise reform agenda,

> Acting Director of Commissioning – What I think is a bit, I don't know whether it is hoodwinking or not, a lot of GPs think that commissioning is the main agenda. Whereas I think the Department of Health thinks that the provision of primary and community care services is the primary agenda.
>
> (*Interview data*)

The avoidance of definitional work as described above is understandable if one considers the possibility that within such a process, power differentials, competing agendas, and even contrasting operating philosophies might become apparent, and in turn create the potential for conflict between local actors. GPs and managers avoided the potentially conflict-ridden task of establishing common understandings of key terms and objectives and establishing norms by essentially "talking past" one another, and tolerating various seeming miscommunications. Actors were able to avoid instigating confrontation while continuing to air competing perspectives, and the incorporation of contrasting definitions, applications and foci actually served to facilitate the continued expression of numerous agendas and perspectives.

Furthermore the "fuzziness" around key terms was sometimes used as a delaying tactic. The Casterdale data include several instances where key ideas or terms were applied in particular ways by actors holding different perspectives of the change process, in which such differences did not seem to hinder communication or co-operation. In these instances, the existence of a variety of understandings of key terms seemed to have facilitated the working process. "The reduction of inequality" was one example of an official policy rhetoric to which participants applied and tolerated contrasting working definitions (accountability, collegial working and clinical governance are further examples for which divergent definitions and applications were successfully employed).

In implementing *The New NHS*, GPs and managers legitimately focused upon various aspects of the key term "inequality": inequalities in resource allocation, inequality of available services, inequality of care given, inequality of health outcomes, or even inequality of life chances. Each perspective falls clearly within the bounds of the policy, and different definitions of terms could be applied depending on both the context and the actors involved. General practitioners were more likely to present an account of inequality of resource allocation, or the availability of services. Managers, on the other hand, emphasized differentials in quality of care or health outcomes that required greater oversight.

Furthermore, among GPs, it served some to apply a restricted definition of "health" when setting out the objective of reducing inequalities through these reform measures. In one instance, a GP whose practice was in a deprived, inner-city area gave the following description of equality.

> If you have got an area that has got the most inequalities in it and inequities – in all of the determinants of health – housing, unemployment, poor education, poor leisure facilities, over-crowding – then the likelihood is that the opportunities for doing something about it now are better than they have ever been because there are funding streams.
>
> (*Interview data*)

This GP presented an image of equality based on the reallocation of resources in favor of those deemed to be most in need. Such an all-inclusive interpretation, as relating to the improvement of a sense of overall well-being, necessitated greater resources being directed towards practices such as his own. Couching such appeals for a greater share of resources within the policy rhetoric was a practical strategy by which to advance his own agenda while avoiding confrontation with GP colleagues.

GPs from more affluent areas simply proposed models of reducing inequalities that built on alternate definitions. Their accounts tended to prioritize equality of the provision of services *within* particular geographic areas or social strata. The following extract is from an interview with one such GP. He frames the reduction of inequalities with reference only to patients from his locale. His focus is also on healthcare, rather than generalized well-being.

> Locally, the sorts of things that we are supposed to concentrate on are areas of deprivation where the healthcare is of a lower standard than elsewhere. Getting good information from public health and the center and concentrate on the sort of areas like smoking in teenage girls, the incidence of heart disease, earlier diagnosis and treatment of cancer, accidents in the home,

> mental health in the community and a couple of those areas are also sort of national priorities and improve those in the patch that we are responsible for.
>
> (*Interview data*)

In this way, these two GPs were able to continue to work together towards reform implementation, while continuing to hold quite different notions of the ultimate objective of reducing inequalities. These findings mirror those from previous research. Light (1999) notes, for example, that when actors often use the term efficiency it is often in confusing and contradictory ways. Likewise, Hughes and Griffiths' (1999) research around the process of setting contracts in NHS hospitals identifies that managers and doctors often concurrently hold different interpretations of key terms (such as "contract"), that are, in turn, accepted and incorporated into the negotiation process. In Casterdale, different applications of key terms were used by GPs and managers to simultaneously promote alternate perspectives. Most actors seemed keen to avoid a situation whereby any forced resolution of the definition of key terms may have led to the breakdown of local co-operation.

"Clinical governance" (essentially the idea of increased monitoring of professional practice either by colleagues or by managers) was another aspect of the reform rhetoric for which actors tolerated imprecise definitions. The idea of clinical governance was engaged within the reform implementation in ways that essentially left its essence unexamined. In particular, the managers avoided providing GPs with a precise definition, and thus GPs were denied the necessary details upon which to mount any effective objection. As the abstract ideas of clinical governance are key to notions of professional accountability, opposition to theoretical proposals was quite difficult to justify. Operational details that would have posed more challenges were generally either largely forgotten or avoided. Once the general notion gained acceptance, resistance against particular measures would prove more difficult. On the other hand, avoiding engaging in a detailed discussion of clinical governance may also have served the GPs. The imprecision around the term may have allowed GPs to delay the implementation of any practical changes. Retaining a certain level of ambiguity around an idea avoids having to deal with problematic details, and the potential resistance of those who might be adversely affected. It may also facilitate framing a consensus in relation to a commonly accepted abstract idea without promoting any actual change. Retaining ambiguity allows one to constantly strive towards a stated goal without fearing the consequences of actual implementation (Meyer & Rowan, 1978).

These data serve not only as examples of instances where actors clearly held different interpretations of key ideas, but also where they accommodated a level of imprecision around terms. Heritage (1984) describes definitions as being continual compromises between generality and specificity. From his perspective, language is "indexical," and its application is based upon practical necessity. There is always an optimal level of specificity for the terms that one uses. These data demonstrate that precise definitions (either distinct or shared) are not always the ultimate objective; sometimes "vague" definitions actually facilitate collaboration.

Manning (1977) identified various ways in which a lack of clarity of terms was actively used by participants to enable better communication. In relation to police work, he constructed a model of six different conceptualizations of the term "major violator."

> If organizations are rule-negotiating contexts, then the shared and non-shared bases for negotiation are of operative importance (Manning, 1977, p. 50).

While operational definitions are effectively used to imply consensus; intractable ambiguity can be used to justify one's behavior, and to associate it with official goals. Thus, the same term can be used to indicate very different symbolic conceptions of goals, while also allowing actionable targets to shift – all within an mutually acceptable framework.

Consensus Strategies: Selective Facilitation

Consensus maintenance was not always controlled unilaterally, but was rather in the hands of particularly influential local actors. More powerful actors were able to direct discussion in such a way as to promote particular outcomes as well as to prevent embarking on a potentially divisive trajectory. Thus, Greatbach and Dingwall's (1989) theory of selective facilitation has considerable salience in relation to this interaction. Greatbach and Dingwall observed that divorce mediators were often able to maintain an image of neutrality while also significantly shaping the outcome of mediation sessions. Mediators steer discussion by selectively facilitating certain comments, and ignoring other less favored options. Likewise, GPs and managers who took leadership roles in the implementation of the reforms were adept at "selective facilitation," using this strategy for controlling the trajectory of reforms. Such tactics were evident from the initial reform implementation meeting.

The initial implementation meeting was open to all local GPs, and was attended by many who would play little further active part in the implementation process. For many of these GPs, this meeting was their first opportunity to express concern about the changing structures, and several expressed dissatisfaction with the decisions taken on their behalf by their GP colleagues. One GP questioned the interim board's authority, and another suggested that the general meeting should take a vote of confidence in the interim board. A show of hands was subsequently conducted, and the result seemed fairly evenly split (from my observation – as well as from the comments of the GPs sitting around me). The GP board members, however, seemed to ignore the somewhat unfavorable outcome of the impromptu vote. The Chair did not take an official count, but instead declared that the decisions taken by the board on behalf of the other GPs to stand. He stated that, "The status of the show of hands was to inform the board, rather than to take a decision." In this way, he justified subsequently disregarding GPs' opposition.

During this large open meeting, one of the GPs (from the floor) also asked interim board members,

> GP – If the existing collective structure becomes lots of smaller structures, will each of these be given (clinical and resource allocation) freedom?

The interim board Chair who did not favor any option of forming several smaller organizations, and essentially ignored the question, instead stating,

> Chair – We may well stay as a single organization. I am not aware of any funds being made available to prepare for (multiple) PCGs.

Other board members supported the Chair's comments by asserting that as the government was looking to save money on administration they were unlikely to put additional money into supporting the formation of numerous local professional structures. This argument was put forward seemingly as a means of dissuading other GPs from pursuing the idea of a totally new organizational structure. While the GP board members did not overtly disagree with potential alternative structures being proposed by their colleagues, they simply overlooked the potentially challenging question by steering discussion towards the problems associated with the scenario outlined in the question.

Later data suggest that the Chair's response to the question of the effects of different structural configurations (as well as the comments from the other Board members) was based largely on an assumption. In fact, the formation of a single, large PCG did not receive either government or local managerial support. In this particular meeting, however, the Chair was able to avoid addressing the question of whether the new PCGs would be given additional resource allocation autonomy by focusing on the possibility that new structures could be avoided.

There were numerous other instances where the opportunity for members to "air" opinions was employed seemingly in order to gain general assent and compliance, without those in power necessarily displaying any clear intention of incorporating expressed opinions into subsequent decisions. One such issue was the configuration of PCGs (how many GP practices would be grouped together to form a PGG – and which practices would be grouped together). In relation to this issue, managers were able to assert their influence over that of the local GPs in various ways during the configuration consultation. The GPs largely supported a single organization model, but this had essentially been covertly rejected as an option by the local managers before the period of consultation began. The option of a single PCG was retained in the official consultation documents and process between managers and GPs, but only discussion on the model preferred by managers was facilitated in meetings.

A Health Authority manager reflected on the decision to include the "single PCG" option within the consultation process.

> Health Authority manager – And a third option was having a single PCG (sighs), which we had sort of discounted early on but there had been a lot of talk with ministers by some of our GPs, and we were advised that we needed to consider it right up until the end or else we could have been taken to judicial review for being prejudicial against it. So, on that advice we included that as an option in the paper. What the paper did was for each option went through the pros and cons for it, and then looked at the level of support that it had. It was obviously easy to talk about the support for this one (the Health Authority model based on local boundaries) because this had been considered as a model itself. So, lots of people had responded.
>
> (*Interview data*)

The manager presented the decision to include the "single PCG" option throughout the configuration consultation as a strategic move. Although the formation of a single PCG was theoretically possible, the lack of consultation in relation to this option meant that there was little evidence of any support, and thus the option could be easily discounted.

Manning's (1977) observation that organizational knowledge is not simply filed for general use, but is rather differentially available depending on one's power, is relevant to the consideration of "selective facilitation" as a means of maintaining surface level agreement, and apparent neutrality. The instances of selective facilitation that were observed in Casterdale were instigated and controlled by those in official positions of power – be they managers or GPs. Selective facilitation served to "smooth the waters" by disabling potentially problematic individuals who did not hold such positions of control.

Consensus Strategies: Managers Working to Establish a Continued Influence

Although professionals are frequently portrayed as protagonists during periods of reform implementation, local managers are also key to directing this process. Managers are particularly influential to the development of a sense of either collaboration or conflict in the local setting. These data suggest that local managers largely worked to establish and support collaboration with GPs – presumably because it furthered their own position within the emerging structure to do so. The newly forming PCGs incorporated complex accountability structures that were challenging for the established roles of both clinicians and managers. The power of different groups in the emerging structures had not yet been determined, and uncertainty was particularly apparent in relation to the continuing role of managerial staff. One manager who was interviewed acknowledged these impending changes.

> Health Authority PCG Implementation Officer – I think that (managers) are sort of having to change their thinking and having to be a support mechanism for PCGs, rather than the governing body.
>
> (*Interview data*)

Thus, managers' co-operation can be interpreted as an effort to establish a new role in terms of providing support for PCGs. Managers' precarious position meant that they could little afford to undermine the power of the newly forming organizations. Health Authority managers did, however, also work to limit the transfer of authority from managers to professionals by conceptualizing themselves as determining the pace with which responsibility and power were to be shifted. Managers also produced accounts of GPs as struggling with their newly acquired workload and responsibilities. During a period when many commentators were posing questions as to what the continued role of local managers might be, it seems to have been valuable for managers to construct accounts in which they were presented as firmly in control over both the pace and direction of change.

> Health Authority Manager – We wouldn't allow them [GPs] to have access to something that we wouldn't give to everybody. Not at least until the first year is out of the way. Because we have to have a gauge of what they can, or think that they can do . . .
>
> We decide what we want to hand down, because we have to be convinced that they are ready to take certain, and we don't want to disenfranchise the Trusts (Secondary Care structures) as well – to put them at risk.
>
> (*Interview data*)

The new PCG structures posed a challenge to the legitimacy of local managers in the realm of decision making. To the extent, however, that managers were able to successfully conceptualize GPs as inexperienced and overburdened in relation to their newly acquired managerial tasks, so managers were able to carve a niche for themselves as essential support mechanisms during the transitional period. Such a position, however, also rested on the ability to demonstrate having a working relationship with GPs.

GPs simultaneously used their status as "apprentices" in management to lessen their perceived responsibility for potentially unpopular or difficult resource allocation decisions. While GP members of PCGs did not produce accounts in which managers were in control of the pace and direction of change, neither did they present themselves as either unready or unwilling to take on additional responsibilities. Any resistance was more readily attributed to professional skepticism, or caution, in relation to central measures, or as an inevitable element of a gradual transfer of authority. GPs demonstrated a willingness to accept a slow pace of power devolution when it occurred alongside an equally slow engagement with accountability for difficult decisions. In this way, confrontation was avoided; the slow pace of change, to a certain extent, suited both managers and GPs.

Consensus Strategies: Calling Upon Worst Case Scenarios

The PCG implementation period was a time of considerable uncertainty and flux. Much of the interaction observed between participants in the implementation meetings, as well as the accounts given in the interviews, dealt with "*what if?*" issues. Predictions as to what might occur were frequent agenda items in the transition meetings, and such predictions often seemed to be used to facilitate the planning process. When "what if?" questions were posed, both managers and GPs called upon extreme examples. Actual experiences rarely seemed pertinent to the construction of possible scenarios. The extreme examples were instead utilized as mechanisms to unite members and to "close ranks," thus avoiding detailing with the "nitty gritty" details of pertinent issues.

The following extracts are instances in which "worst case scenarios" were provided in response to requests for predictions about the future of PCGs. The provision of the worst possible outcomes enabled actors to avoid discussing the details of more likely scenarios. The following is taken from an implementation meeting in which GPs discussed options with regards to the proposed structural changes.

> GP 1 – Should we as a group require the Chief Executive (top local manager) to assure us that GPs will need to be involved at the time of decision making?
>
> Chair – They can say, 'Sod off. You are a subcommittee and we will do what we want.'
>
> (*Observational data*)

My data reveal little support for the notion that such overt confrontation and assertion of authority on the part of any manager. Another GP called upon the worst possible outcome when one PCG chose to frame the issue of increased lay involvement in decision making in relation to the possibility of having to sanction failing GP practices.

(Extract from an implementation meeting among GPs only where the inclusion of a lay representative in decision making was being discussed.)

GP – I wouldn't be happy for them to be here in cases where we are dealing with failing practices. In such cases, the less people in the room the better.

(*Observational data*)

Until this point, no case of a failing practice had ever come before any of the newly forming structures, nor had there been any detailed discussion as to how such an issue might be dealt with. Yet, in this PCG meeting, the discussion about including a lay representative on the board was dominated by a very negative portrayal of the possibility that this would be prohibitive to possible future professional regulation. The possibility of other pros and cons were not discussed because the available time (and energy!) was spent in regards to this one seemingly peripheral issue.

DISCUSSION

Evaluations of health care reform measures often conclude that efforts fail to bring about desired change. In turn, such failures are sometimes attributed either to staunch professional resistance or to the impact of the competing agendas of local professionals and mangers. These data, however, provide little evidence to suggest that the reforms in question were being actively and openly opposed at the local level, although the initiatives were receiving considerable critical attention. Instead, local implementation seemed to involve most actors acknowledging the key reform objectives, and subsequently working to incorporate the reforms in some way into existing working practices, while also maintaining (and possibly strengthening), local working structures. Systems of consensus and co-operation were built between local management and GPs, with the seeming result of facilitating the different stakeholders' continual pursuit of their own agendas.

It may have been that local implementation was shaped by GPs' considerable cultural authority to limit the impact of any proposed changes (Griffiths & Hughes, 1999). Indeed, Harrison et al. (1992) suggest that the cultural authority granted to doctors undermines the power of other groups without requiring overt confrontation. The cultural dominance of doctors may be such that other actors are unlikely to consider the possibility of introducing initiatives that might challenge the authority of the medical profession, with professional discourse being so embedded as to actually shape what is deemed "thinkable" or "doable" (Dingwall & Strong, 1985). The absence of overt confrontation between managers and GPs in Casterdale might not, therefore, reflect a simple "reform inertia," but rather the ability of the GPs to protect the status quo that they perceive to have served both themselves and their patients. This explanation does not, however, fit particularly easily with the Casterdale data. Local actors did not totally ignore the reform initiatives, or the ideas behind them. While the cultural authority of doctors is certainly apparent, these data suggest that the GPs had to carefully negotiate the implementation of reform initiatives with local managers. Such tasks involved considerable efforts at the local level to peacefully manage competing agendas for reform interpretation and implementation.

In turn, the managers were seemingly adept at using the power void that was created by the changing structures to reinforce their own influence over local systems. Both GPs and managers seemed to appreciate that they stood to lose more by not being involved in shaping the newly forming structures. On the other hand, both groups also worked to avoid the perception of "leading the charge" of change on potentially controversial issues. One of the shortcomings of the traditional professional/managerial "conflict" model is that it is based upon an assumption of a single hierarchical system of control, in which one group will necessarily dominate all others. Such models have typically presented professional and bureaucratic models of control as oppositional, and periods of reform as opportunities in which existing power structures might be challenged. Business enterprises are, however, increasingly likely to be organized according to "network" principles that emphasize collaboration and organizational interdependency, rather than around vertical (hierarchical) integration (Flynn, Pickard & Williams, 1994). Internal networks and coalitions are also becoming increasingly important, such that they can be used to pursue identified common interests (Bacharach & Lawler, 1980).

Network systems are particularly prevalent where relationships are long-term and involve high levels of co-operation around work, the pooling of resources, burden sharing and reciprocity. In "clan" (Ouchi, 1991) or "network" systems, performance ambiguity is often tolerated, and goal conflicts between actors minimized, in order to protect the complex, valuable and long-term existing relationships. This organizational structure resonates strongly with the Casterdale data. Furthermore, these models of networked relationships reflect a prioritization of Durkheimian principles of "organic solidarity," in which individuals make different, but interrelated contributions to a collective product.

During this reform implementation, both GPs and managers were motivated to make the reforms successful. The background of "near constant reform" within the NHS provided actors with a vested interest in protecting existing and newly forming social networks, regardless of the particular outcome of the present reforms. The creation of an atmosphere of consensus and co-operation in relation to the reform initiatives in Casterdale might be understood as a particular strategic action on the part of local actors. GPs and managers were able to incorporate ideas from the reforms that supported their own agendas, while also protecting the important local, working networks that their experience tells them are likely to outlast the particular present initiatives.

SUMMARY

Ham (1992, p. 36) proposes that it is vital to consider negotiation and bargaining within the policy community in order to better understand its detailed processes. Failure to consider the detailed application of policy can lead to simplistic conclusions that policies have fallen short of their goals, rather than appreciating that it is the role of local actors to shape goals into acceptable and workable structures. Influential actors in the local realm (particularly the medical profession) have tended to have a sufficiently strong power base to limit the impact of reform measures (Ashburner, 1995). Whereas previous models have tended to conceptualize the power distribution between doctors and managers as an isolated "zero-sum" games (Klein, 1995), I suggest that actors appreciate the importance of the protecting the local network. By and large, the implementation process for PCGs (between GPs and local managers) is one in which different local stakeholders employed a variety of techniques and mechanisms to facilitate consensus and collaboration wherever possible. I have suggested that this is because both GPs and management recognized considerable potential within such consensus through which they might pursue their own agendas.

NOTE

1. At the time that this research was conducted the Health Authority was the local unit through which most health care resources were administered. The health authority consists of managers and administrators whose job it is to oversee the delivery of care to the local population by medical professionals.

REFERENCES

Ashburner, L. (1995). The role of clinicians in the management of the NHS. In: J. Leopold, I. Glover & M. Hughes (Eds), *Beyond Reason? The NHS and the Limits of Management*. Aldershot: Avebury.

Bacharach, S., & Lawler, E. (1980). *Power and politics in organizations*. San Francisco: Jossey Press.

Calnan, M., & Gabe, J. (1991). Recent developments in general practice: A sociological analysis. In: M. Calnan, J. Gabe & M. Bury (Eds), *The Sociology of the Health Service*. London: Routledge.

Clarke, J., Cochrane, A., & McLaughlin, E. (1994). Mission accomplished or unfinished? The impact of managerialization. In: J. Clarke, A. Cochrane & E. McLaughlin (Eds), *Managing Social Policy*. London: Sage.

Coombs, R. W. (1987). Accounting for the control of doctors. *Accounting Organizations and Society*, *12*, 389–404.

Deal, T., & Kennedy, A. (1988). *Corporate culture: The rites and rituals of corporate life*. London: Penguin Books.

Department of Health (1997). *The new NHS: Modern, dependable*. London: HMSO.

Dingwall, R., & Strong, P. (1985). The interactional study of organizations: A critique and reformulation. *Urban Life*, *14*, 205–231.

Dopson, S. (1996). Doctors in management: A challenge to established debates. In: J. Leopold, I. Glover & M. Hughes (Eds), *Beyond Reason: The National Health Service and the Limits of Management*. Aldershot: Avebury.

Drummond, M. (1998). Evidence-based medicine and cost-effectiveness: Uneasy bedfellows? *Evidence Based Medicine*. http://www.acponline.org/journals/ebm/sepoct98/bedfell.htm

Ferlie, E. (1997). Large-scale organizational and managerial change in health care: A review of the literature. *Journal of Health Services Research and Policy*, *2*(3), 180–189.

FitzGerald, L. (1996). Clinical management: The impact of a changing context in a changing profession. In: J. Leopold, I. Glover & M. Hughes (Eds), *Beyond Reason: The National Health Service and the Limits of Management*. Aldershot: Avebury.

Flynn, R. (1992). *Structures of control in health management*. London: Routledge.

Flynn, R., Pickard, S., & Williams, G. (1994). Contracting and the Quasi-market in community health services. *Journal of Social Policy*, *24*, 529–550.

Freidson, E. (1975). *Doctoring together: A study of professional social control*. Chicago: University of Chicago Press.

Freidson, E. (1993). How dominant are the professions? In: F. Hafferty & J. B. McKinlay (Eds), *The Changing Medical Profession: An International Perspective*. New York: Oxford University Press.

Greatbach, D., & Dingwall, R. (1989). Selective facilitation: Some preliminary observations on a strategy used by divorce mediators. *Law and Society Review*, *23*, 613–641.

Greenfield, S., & Nayak, A. (1996). A management role for the general practitioner? In: J. Leopold, I. Glover & M. Hughes (Eds), *Beyond Reason: The National Health Service and the Limits of Management*. Aldershot: Avebury.

Griffiths, L., & Hughes, D. (1999). Talking contracts and taking care: Managers and professionals in the British National Health Service internal market. *Social Science and Medicine*, *4*, 1–13.

Ham, C. (1992). *Health policy in Britain: The politics and organisation of the National Health Service* (3rd ed.). Houndsmills: Macmillan.

Ham, C. (1998). The new NHS: Commentaries on the white paper. Financing the NHS. *British Medical Journal*, *316*, 7126.

Harrison, S., Hunter, D., Marnoch, G., & Pollitt, C. (1992). *Just managing: Power and culture in the National Health Service*. Basingstoke: Macmillan.

Haug, M. (1973). Deprofessionalisation: An alternative hypothesis for the future. *Sociological Review Monograph*, *20*, 195–211.

Heritage, J. (1984). *Garfinkel and ethnomethodology*. Cambridge: Polity Press.

Honigsbaum, F. (1979). *The division in British medicine: A history of the separation of general practice from hospital care 1911–1968*. London: Kogan Page.

Hughes, E. C. (1964). Professions in transition. In: E. C. Hughes (Ed.), *Men and Their Work*. London: Free Press of Glencoe.
Hughes, D., & Allen, D. (1993). *Inside the black box: Obstacles to change in the modern hospital*. King's Fund and Milbank Memorial Fund. Joint Health Policy Review.
Hughes, D., & Griffiths, L. (1999). On penalties and the "patient's charter": Centralism vs. decentralised governance in the NHS. *Sociology of Health and Illness*, *21*, 71–94.
Klein, R. (1995). Big bang health care reform – does it work? The case of Britain's 1991 National Health Service Reforms. *The Milbank Quarterly*, *73*, 299–333.
Larson, S. (1977). *The rise of professionalism: A sociological analysis*. London: University of California Press.
Light, D. (1993). Countervailing power: The changing character of the medical profession in the U.S. In: F. Hafferty & J. B. McKinlay (Eds), *The Changing Medical Profession: An International Perspective*. New York: Oxford University Press.
Light, D. (1999). The sociological character of health care markets. In: G. Albrecht, R. Fitzpatrick & S. Scrimshaw (Eds), *Handbook of Social Studies in Health and Medicine*. London & San Francisco: Sage.
Manning, P. (1977). Rules in organizational context: Narcotics law enforcement in two settings. *Sociological Quarterly*, *18*, 46–63.
March, J. G., & Olsen, J. P. (1983). Organization political life: What administrative reorganization tells us about government. *The American Political Science Review*, 280–296.
Meyer, J., & Rowan, B. (1978). The structure of educational organizations. In: M. Meyer et al. (Eds), *Environments and Organizations*. San Francisco: Jossey-Bass.
McKinlay, J. B., & Arches, J. (1985). Towards a proletarianisation of physicians. *International Journal of Health Services*, *15*, 161–195.
Miller, G. (1970). Professionals in bureaucracy: Alienation among industrial scientists and engineers. In: O. Grusky & G. Miller (Eds), *The Sociology of Organizations*. London: Collier Macmillan Publishers.
Ouchi, W. G. (1991). Markets, bureaucracies and clans. In: G. Thompson et al. (Eds), *Markets, Hierarchies and Networks*. London: Sage.
Rose, N., & Miller, P. (1992). Political power beyond the state: Problematics of government. *British Journal of Sociology*, *43*, 173–205.
Robinson, R., & Steiner, A. (1998). *Managed health care*. Buckingham: Open University Press.

SOCIAL CAPITAL, HEALTH STATUS, AND HEALTH SERVICES USE AMONG OLDER WOMEN IN ALMATY, KAZAKHSTAN

Thomas T. H. Wan and Blossom Y. J. Lin

ABSTRACT

Understanding the determinants of health services use is essential for planning for effective services, particularly health care policies in a newly independent state, the Republic of Kazakhstan. The main purpose of this study is to examine the relative importance of social capital factors in affecting the variation in health status and use of health services, using structural equation modeling. The results show that health status is a strong predictor of health services use when the effect of social capital is held constant; and that social capital is directly linked with health status.

INTRODUCTION

The Republic of Kazakhstan became an independent state after the break-up of the Soviet Union in 1992. The country has experienced a difficult transition from autocratic control to political democracy along with a free market economy. Its health care system is under-financed and has inadequate accessibility and quality of care.

Reorganizing Health Care Delivery Systems: Problems of Managed Care and Other Models of Health Care Delivery
Research in the Sociology of Health Care, Volume 21, 163–180

ISSN: 0275-4959/doi:10.1016/S0275-4959(03)21009-7

The crude death rate, which had dropped substantially in the previous decade, has now increased from 7.7 per thousand in 1990 to 10.5 per thousand in 1995. Standardized mortality rates of cardiovascular diseases and neoplasm have risen among the highest in the WHO's European Region. These indicators reveal the urgency of improving the health status of the population and reforming the health care system.

The change from a communistic to a democratic state will impact on public and personal health of the people in Kazakhstan. However, little is known about how personal and societal factors influence the use of health services in this newly independent state. The growing impetus from market forces in Kazakhstan over the past few years may be reducing the use of inpatient services and shifting demand to ambulatory care services. If the relationship there between ambulatory care and inpatient care becomes a substitutive rather than a complementary one, the implications for health services planning and demand management will be significant. Moreover, the transformation of social medicine from a single tier to a multi-tier system presents the picture of a relatively free choice of medical care, contingent however on the ability to pay for it. Disparities between the health care for the poor and that for the rich were observed in Almaty, the largest city in Kazakhstan. However, official utilization statistics compiled from administrative and survey sources on health services use are not available. Thus, a community-based survey of personal and societal factors affecting an individual's use of health services is imperative in setting health service priorities.

Andersen's behavioral system model is widely used as a conceptual framework for exploring health services use (Andersen, 1995; Andersen & Newman, 1973; Wan, 1989; Wolinsky & Johnson, 1991). In brief, the behavioral system model takes both societal and individual determinants into account in its view of health services use, and focuses on evaluating the various individual determinants. The individual determinants are generic factors in an individual's background that fall into three distinctive groups: predisposing characteristics, enabling resources, and health status (i.e. need for care). Predisposing characteristics are the personal attributes that may predispose an individual to seek or not to seek health services; they include demographic, social structural, and attitude-belief variables. The enabling component comprises factors that may either enhance or impede an individual's access to health services. Health status (or need for care) is the most immediate cause of services use. For services to be used, an individual must perceive the necessity of seeking care for illness, and clinical evaluation must confirm the necessity. The purpose of this study is to examine the effect of the social structure as specified in social capital on health services use by older women in Almaty, Kazakhstan, assessing those effects both directly and indirectly through women's health status.

RELATED RESEARCH AND TESTABLE HYPOTHESES

For almost half a century, social epidemiologists have been researching the social determinants of people's health and illness. Although there is no question about the importance of clinical, biological, and behavioral risk factors for illness, it is posited that individual behaviors in using health services are conditioned by social environments through norms, enforcing patterns of social control, opportunity constraints, and stress occurrence (Berkman & Kawachi, 2000).

Social capital is considered a major social force and concept influencing people's health and health care. The concept of social capital is defined as those features of social structures that function as resources and facilitate cooperation for mutual benefit (Coleman, 1990; Putnam, 1993), in addition to the concept of "capital" in mercantile exchange, familiar from economic theories (Bourdieu, 1986). From an ecological approach, social capital can be conceptualized as social cohesion, mutual trust, extent of obligations, norms and effective sanctions, commitment, information channels, perceived reciprocity and civic engagement (Berkman & Kawachi, 2000; Coleman, 1990; Putnam, 1993, 1995). The concept of social capital has been applied in fields studying families and youth behavior, education, work and organizations, democracy and governance, economic development and public health (Berkman & Kawachi, 2000). In public health, recent studies have examined the relationships of social capital to mortality (Kawachi et al., 1997; Lochner et al., 2002), to violent crime (Kennedy et al., 1998; Sampsom et al., 1997), to teen birth rate (Gold, 2002), to self-rated health status (Kawachi et al., 1999; Veenstra, 2002), and to binge drinking (Weitzman & Kawachi, 2000). As done by Hendryx et al. (2002), area-level measures of social capital were used to analyze the association between social capital and health care. They suggest that higher social capital might be associated with better access to care because of improved community accountability mechanisms. Several critical points about the causal effect of social capital on health remain unexplored. First, most studies focusing on social capital in the field of health are devoted largely to health-related indicators such as mortality and self-rated health status, rather than examining health services utilization. Incorporating the social context (e.g. social capital) in analyses of behavior could improve our understanding of the realm of individual choice – as, for example, in this study's examination of health services use. Second, in terms of measurement, little has been done to infuse a variety of personal perceptions of social capital, such as of trust, reciprocity, civic engagement and so on into the conceptual domains of social capital. In other words, social capital as a multidimensional construct measurable from personal views on trust, social involvement, and reciprocity should be considered integral to the social structure that may influence health and health care. The use of proxy or crude ecological indicators of social structure may not adequately reflect personal perceptions of social pathogenesis of disparities in health and health care.

This study conducted with older women residing in Almaty, Kazakhstan examined the pattern of their health services use. Social capital, with three first-order theoretical factors (trust, reciprocity and civic engagement) was conceptualized and validated in a three-factor measurement model. The relative importance of each domain of social capital for health status and health services use was examined. It is postulated that social capital influences both health status and health services use. The major hypotheses of this investigation are as follows:

Hypothesis H1. Social capital exerts both direct and indirect effects on use of health services.

Hypothesis H2. Holding social capital variables constant, health status is inversely related to use of health services.

METHODS

Data Sources and Sample Selection

The primary purpose of the interview survey was to gather information about breast cancer screening and preventive practices among older women. A purposive sample of 500 women stratified by ethnic groups was designed to collect personal information from 100 zip code areas. There were 252 Korean Kazakhs and 248 non-Korean Kazakhs in the study sample. The sample was selected among women if they met the following criteria:

(1) Having lived in Almaty, Kazakhstan for more than a year.
(2) Not an inpatient of a hospital at the time of the survey.
(3) Forty-five years of age or older in November 2001.

Four faculty members of the Kazakhstan School of Public Health conducted personal interviews. They were instructed by the principal investigator on survey research and design, as part of the Richmond-Almaty Collaboration Program sponsored by the American International Health Alliance.

Survey Instrument

The survey booklet, containing a total of 110 questions, requires approximately 40 minutes to complete in a personal interview. The questionnaire has questions about demographic, social, economic, cultural, preventive health attitudes, health functioning, knowledge about breast cancer, perceived barriers to preventive screening for breast cancer, health services use, and social capital. The English version of the interview questionnaire was translated into Russian, the language most used in Kazakhstan. To assess the validity of the translation, two bilingual individuals examined the conformity of meaning between each item of the questionnaire and its Russian translation. A revised Russian version of the questionnaire was then back translated by a bilingual Kazakhstan instructor to validate the translation. Before conducting the survey, we pre-tested the questionnaire with five interviewees to improve its validity.

Social Capital Characteristics: Exogenous Variables

Following the previous studies by Kawachi et al. (1997, 1999) and Lochner et al. (2002), social capital was conceptualized as three related constructs including trust, social involvement, and reciprocity/benefits as perceived by the respondents. Each question was scored by a four-point scale ranging from 1 (lowest) to 4 (highest). Trust is defined by the extent to which a respondent feels comfortable and trusting others in the community (TRUSTCOM), considers that people or public programs are helpful (HELPCOM), and shares a common interest with the community (COMCOM). Social involvement is a similar concept of associational membership (Kawachi et al., 1997, 1999; Lochner et al., 2002), engagement in politics (Putnam, 1995) and civic participation (Veenstra, 2002). Social involvement is measured by the extent that a respondent actively participates in community affairs (ACTINVOL) and the frequency of attending social activities in the community (ATTENDEN). Reciprocity is defined by the extent to which the respondent reports having benefited from participating in public activities or programs (BENEFIT), having learned about specific medical services available in the community (LEARN), and sharing a sense of common goals and interests in promoting health (SHARE) with others in the community.

Health Status: An Endogenous Variable

Among different approaches to measuring population health, the use of the Short-Form Health Survey (abbreviated as SF-12) has become popular for both research and administration. Briefly described, the SF-12 Health Survey measures physical and mental health functioning: general health (GHscore), physical functioning (PFscore), role limitations due to physical health problems (RPscore), bodily pain

(BPscore), role limitations due to emotional problems (REscore), and mental health (MHscore). A more detailed description of the SF-12 Health Survey is provided in the manual (Ware et al., 1993, 1995). Following the same computational algorithms specified by SF-12, we calculated each of the six indicator scores. Because our interest in assessing the measurement model of six health-status indicators, we did not generate the aggregate scores of physical and mental health.

Health status is a latent construct that can be reflected by indicators of physical and mental functioning in the study population. A measurement model of six indicators of health status was developed and validated by confirmatory factor analysis.

Utilization of Health Services: An Endogenous Variable

In this study, health services use is considered a latent construct to be measured by six related indicators of services use: numbers of physician visits (PHYVISIT), pharmacist visits (PHAMVIS), dental visits (DENTVISI), alternative medicine visits (ALTNVISI), hospital visits (HOSVISIT), and outpatient visits (OUTPVISI). A measurement model of health services use was also evaluated by confirmatory factor analysis.

Statistical Modeling Approaches and Techniques

In order to test the relative importance of different aspects of social capital on health status and health services use, the direct effect of each social capital dimension was examined by structural equation modeling. In addition, the indirect effects of social capital variables via health status on use of health services were studied. The statistical analysis is based mainly on structural equation modeling (SEM), with the aid of the AMOS 4.0 software program (Arbuckle & Wothke, 1997). A structural equation model evaluates whether the proposed causal relationship is consistent with the actual patterns found among variables in the empirical data.

To achieve that purpose, the SEM uses a two-step process: the measurement model and the structural equation model. Put briefly, the measurement model specifies how the latent (unobserved) variables or hypothetical constructs are measured in terms of the observed variables. Expressed in mathematics, the observed variables and unobserved constructs are linked by one of two factor equations, as follows:

$$X = \Lambda_x \xi + \delta,$$

wherein Λ_x is a matrix of the loadings of the x's on the ξ's; ξ is a vector of latent variables; δ is a vector of unique factors or errors in measurement that affect the x's; X is a vector of observed exogenous variables.

$$Y = \Lambda_y \eta + \varepsilon,$$

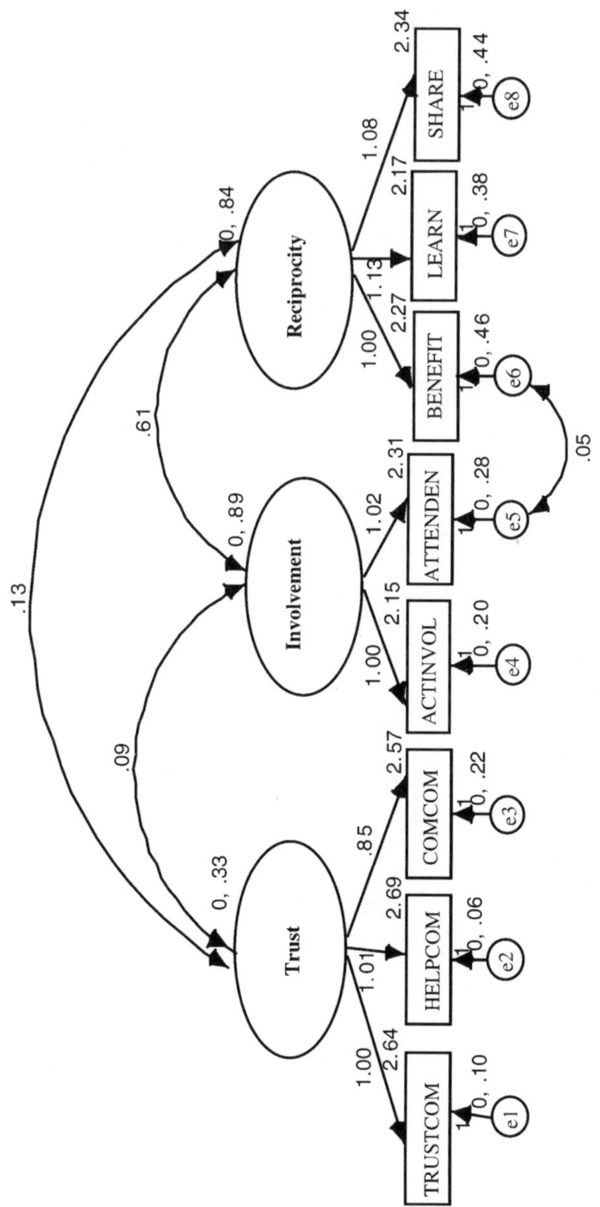

Fig. 1. A Measurement Model of Social Capital.

wherein, Λ_y is a matrix of the loadings of the y's on the η's; η is a vector of latent variables; ε is a vector of unique factors or errors in measurement that affect the y's; Y is a vector of observed endogenous variables.

Based on results derived from the measurement models, the structural equation model specifies the causal relationships among the exogenous and endogenous variables, and describes the amount of unexplained variances among them. A mathematical expression of the structural equation model is specified as follows:

$$\eta = B\eta + \Gamma\xi + \varsigma,$$

wherein η is a vector of endogenous variables; B is a matrix of coefficients relating the endogenous variables to one another; Γ is a matrix of coefficients relating the exogenous variables to the endogenous variables; ξ is a vector of exogenous variables; ς is a vector of errors in an equation.

The structural equation model for this study is diagrammatically illustrated in Fig. 1. After model specification is performed, the assessment of model fit is undertaken to ensure the appropriate interpretation of the theoretical framework. The criteria for assessment of fit include examination of the solution, measures of overall fit, and detailed assessment of fit (Bollen, 1989; Jöreskog & Sörbom, 1989; Wan, 2002). In the first step, parameter estimates with the right sign and size, standard errors within reasonable ranges, correlations of parameter estimates, and squared multiple correlations were commonly used to check for the appropriateness of each variable. In the second step, the overall model fit is evaluated to see how well the specified model fits the data. The indicators adopted here are chi-square value (χ^2), the Tucker-Lewis (1973) index (TLI), the Bentler and Bonett (1980) normed fit index (NFI), and root mean squared error of approximation (RMSEA). The third step is to determine the possible sources of the lack of fit. The commonly used indicators are modification indices, which show the extent to which the model fit could be improved by adding certain constraints between variables. Briefly, chi-square ratio (χ^2/df) is a measure of goodness-of-fit while taking into account the degrees of freedom available. RMSEA (root mean squared error of approximation) measures the degree of model adequacy based on population discrepancy in relation to degrees of freedom; a value less than 0.05 (or 0.08) is acceptable.

RESULTS

Descriptive Statistics of the Study Sample

A purposive sample of 500 women (aged 40–78) was selected from 100 zip code areas in Almaty, Kazakhstan. Because of our original interest in exploring cultural

Table 1. Selected Demographic and Health Characteristics of the Study Sample ($N = 500$ Women) by Ethnic Group.

Individual Characteristics	Mean or Percentage			
	Total ($N = 500$)	Koreans ($N = 252$)	Non-Koreans ($N = 248$)	F-Value (Difference Between Two Groups)
Age (mean)	51.30	51.36	51.23	0.03
Marital status (% married)	65.20	67.20	63.20	0.88
Years of education	14.28	13.98	14.58	1.40
Employment status (% employed)	68.00	65.60	70.40	1.32
Physical health score (mean)	46.13	47.07	45.10	12.49*
Annual physician visits (mean)	2.47	1.63	3.33	6.27*
Visits to alternative medicine provider (mean)	0.62	0.56	0.67	0.32
Dental visits (mean)	1.67	1.26	2.08	1.68
Visits to pharmacists (mean)	3.98	4.04	3.93	0.20
Outpatient visits (mean)	1.93	1.37	2.50	5.84*
Hospital visits (mean)	0.53	0.36	0.71	5.67*

*Statistically significant at 0.05 or lower level.

influences on preventive health practice and behavior, an over-sampled Korean group ($N = 252$) was selected. Comparative statistics for major demographic and health characteristics of the study sample are presented in Table 1. The average age of the study sample was slightly above 51. The majority of the subjects were married. The years of education ranged from 3 to 30, with a mean of 14.3. The sample had a relatively large proportion of the employed (68%). No statistically significant differences were observed in these demographic characteristics between Korean and non-Korean Kazakhs in the sample.

The study sample is a very healthy group, with an average of 46.13 on the physical health score of Sf-12. Among the total study subjects, the average number of annual visits of health providers is 2.5 for physicians, 0.6 for alternative medicine providers, 1.7 for dentists, 4 for pharmacists, and 1.9 for outpatient care. The Korean Kazakhs having a higher average physical health score had visited physicians less frequently, made fewer outpatient visits, and had fewer hospitalizations than the non-Korean Kazakhs had ($p < 0.05$) shown in Table 1.

The consistency of measurement variables within each factor dimension of the social capital construct has been evaluated. The reliability measure (alpha coefficient) is 0.87 for the trust domain, with three indicators; 0.88 for the involvement domain, with two indicators; and 0.87 for the reciprocity or social benefit domain, with three indicators. A measurement model with three related factors of social capital is presented in Fig. 1. Standardized parameter estimates are shown. A moderately high association is observed between involvement and reciprocity ($r = 0.61$). A statistically significant positive relationship of trust to two other domains is also observed, although the correlation is weak. The confirmatory factor analysis reveals satisfactory fit statistics ($\chi^2 = 55.767$, df $= 16, p = 0.000$, TLI $= 0.992$, NFI $= 0.995$, RMSEA $= 0.071$).

Health status is measured by six indicators from the health survey. The confirmatory factor analysis shows that these indicators are positively associated with the latent variable (health status); factor loadings range from 0.51 to 0.82. The fit statistics of this measurement model are very satisfactory ($\chi^2 = 13.334$, df $= 8, p = 0.101$, TLI $= 0.998$, NFI $= 0.998$, RMSEA $= 0.037$) (Fig. 2).

The measurement model of health services use consists of six related indicators. In this model, correlated error terms of the indicators are included. The confirmatory factor analysis reveals that this model fits the data very well, although the factor loading of one indicator (hospital visits) is relatively weak, though statistically significant, compared with other indicators. The fit statistics show that the measurement model of health services use is satisfactory ($\chi^2 = 10.215$, df $= 5, p = 0.069$, TLI $= 0.990$, NFI $= 0.993$, RMSEA $= 0.046$) (Fig. 3).

Social Capital as Predictors of Health Status and Health Services Use

Social capital, the first-order factor consisting of three related domains (trust, involvement, and reciprocity), was postulated to have a direct influence on health

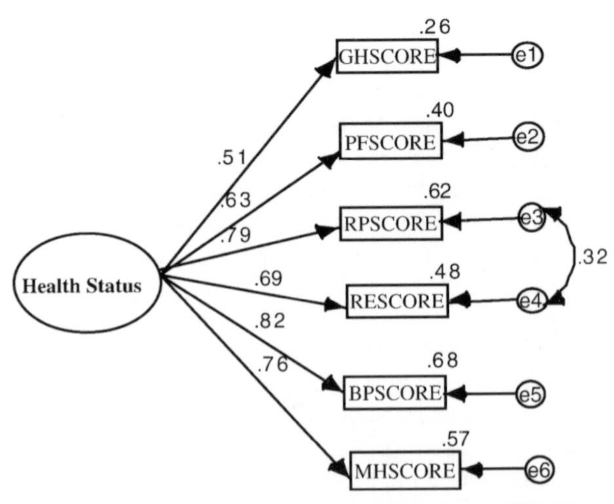

Fig. 2. A Measurement Model of Health Status (SF-12).

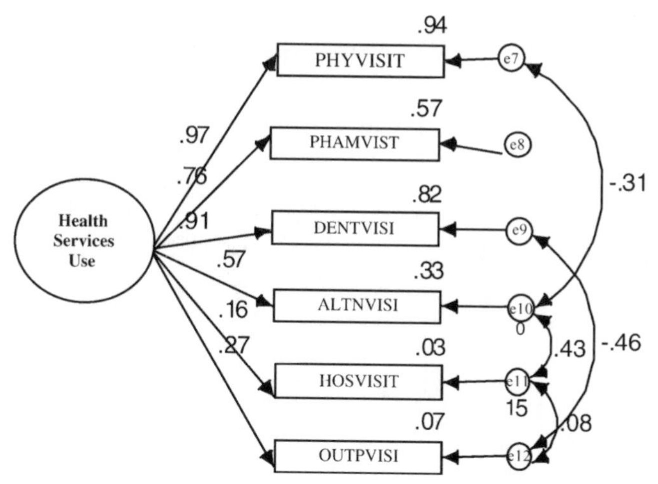

Fig. 3. A Measurement Model of Health Services Use.

Table 2. Maximum Likelihood Estimates for the Structura Equation Models of Health Services Use.

Latent Variable	Standardized Regression Coefficients		
	Proposed Model		Revised Model (Mediating Effect Model)
	Health Status	Health Services Use	Health Services Use
Social capital factors			
Trust	0.148**	0.026	0.147*
Involvement	0.122	0.090	0.118
Reciprocity	0.004	−0.067	0.007
Health status	–[a]	−0.262***	−0.251***
Goodness of fit indices			
χ^2		305.357	307.069
Degrees of freedom		155.000	158.000
Probability		0.000	0.000
TLI		0.990	0.990
NFI		0.985	0.985
RMSEA		0.044	0.043
Hoelter (0.05 level)		303.000	307.000

[a] Not included in the equation.
*0.05 level of significance (two-tailed).
**0.01 level of significance (two-tailed).
***0.001 level of significance (two-tailed).

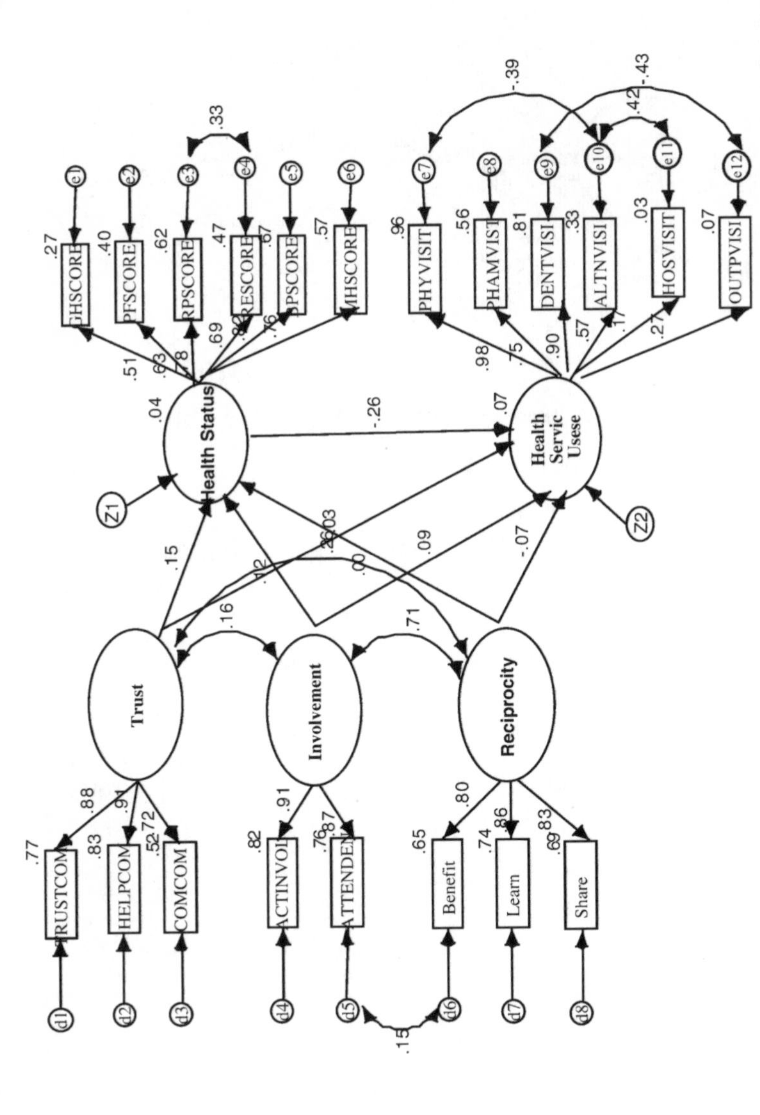

Fig. 4. Social Capital, Health Status and Health Services Use: A Proposed Model of Direct and Indirect Effects of Social Capital.

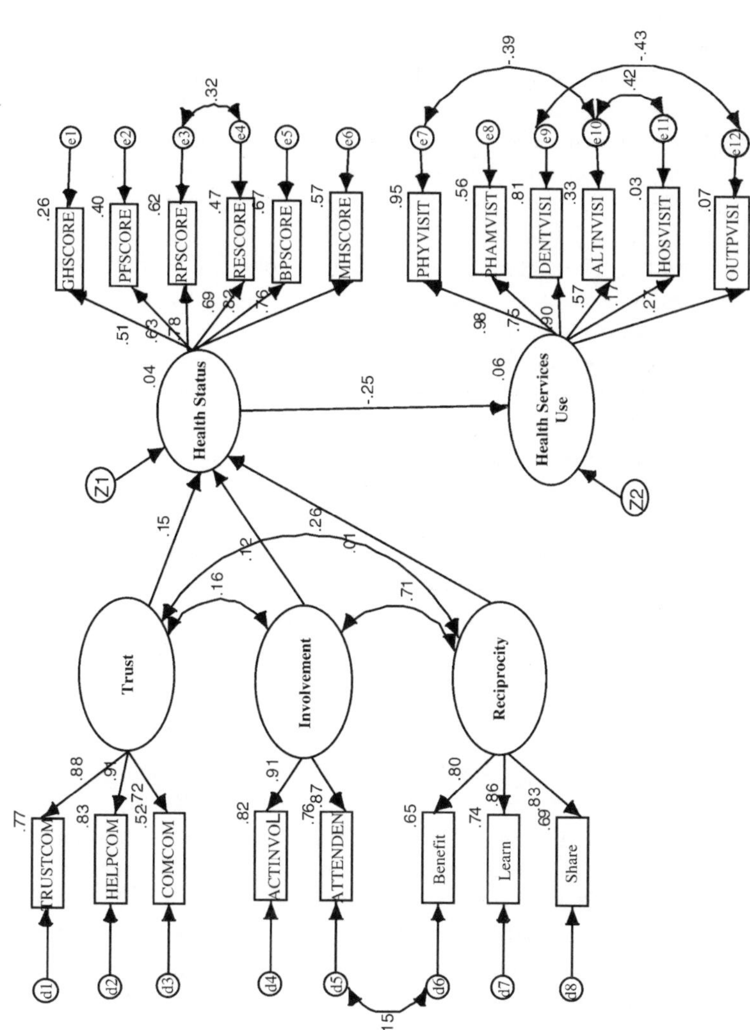

Fig. 5. Social Capital, Health Services Use: A Mediating Effect Model.

status and use of health services. A covariance structure model is presented in Fig. 4. Estimated standardized parameters appear in Table 2. Trust is positively related to health status (gamma = 0.15, with a critical value of 2.77). The two other domains of social capital, involvement and reciprocity, have no significant influence on health status. Social capital exerts no direct influence on use of health services, as shown in Table 2. A statistically significant inverse relationship is found between health status and health services use (beta = −0.26); better health status results in less health services use. Overall, the fit statistics of this covariance structure model show that the proposed model, assuming a direct effect of social capital factors on health status and health services use, is only partially supported by the data ($\chi^2 = 305.357$, df = 155, $p = 0.000$, TLI = 0.990, NFI = 0.985, RMSEA = 0.044). Thus, Hypothesis 1 (social capital exerts both direct and indirect effects on use of health services use) is only partially confirmed: social capital exerts no direct effect on health services use.

From the results presented in Table 2, we find that social capital exerts a direct influence on health status, but has no direct influence on health services use. Thus, the model was revised to assume that health status is a mediating factor between social capital and health services use (social capital → health status → health services use). The results are presented in Fig. 5 and Table 2.

The mediating effect of health status on health services use is found in Fig. 5. As expected, social capital does not directly influence health services use, but exerts an indirect effect on health services use via health status. Thus, Hypothesis 2 is supported. Overall evaluation of the model fit remained the same, with $\chi^2 = 307.069$, df = 158, $p = 0.000$, TLI = 0.990, NFI = 0.985, RMSEA = 0.043.

DISCUSSION

Among a variety of studies on determinants of health services utilization, health status has been the main point of interest. Some studies employing Andersen's behavioral system model have found that need factors, especially self-rated health status, tend to affect health services use most directly. People with lower self-rated health are likely to have more physician visits. More hospitalizations, longer lengths of stay, and higher total charges also are associated with individuals' perceptions of poorer health status. Katz et al. (1996), comparing hospital use in Ontario and in the U.S., also found that across all income groups and in both countries, persons with fair or poor health have about four times the number of admissions than those who report excellent, very good, or good health (41.6 vs. 11.5 admissions per 100 person-years).

Another study (Stump et al., 1995), investigating physician utilization among older adults between 1984 and 1990, found that some changes in functional status are highly associated with changes in use of physician services. That is, when functional status (measured by ADL) declines, the number of physician visits (use) increases; however, improvement in functional status has little effect on physician visits. On the other hand, in one review of 42 empirical studies, Epstein et al. (1988) found that, among the five categories of predictors (perceived health status, functional health status, prior utilization, clinical descriptors, and sociodemographic characteristics), prior utilization ranked first, followed by perceived health status.

In this study, the finding that health status, a latent construct measured by six sub-scales, exerts a statistically significant influence on health services use is consistent with the conventional perception of health care: that in most situations, people residing in the community seek health care services only when their poor health forces them to do so. This pattern not only explains why health status is a dominant predictor of health services use, but also illustrates the phenomenon that health status is pertinently represented by those health concepts highly related to mobility (e.g. physical functioning, bodily pain) and activities (i.e. social functioning), as opposed to cognitive or affective perceptions.

Social capital as an indirect rather than direct effect on health services use is a new finding in health services research. Social capital appears to be a contextual factor that only indirectly affects the variation in health services use through health status. However, this study found that one factor, trust, of the social capital construct does exert a direct influence on health status. This specific finding is unique and should be further explored to clarify how this social context influences personal perceptions of health. Social capital factors, as demonstrated by many other studies, show little statistical power in explaining the variation in use of health services when health status is accounted for. Indeed, although it has been argued that social factors such as social capital may have differing influences on personal health, this study shows that the direct effects of two social capital factors (involvement and reciprocity) on use of health services are not statistically significant. An alternative analytic approach is to aggregate personal perceptions of social capital as ecological variables so the relationship between social capital and health services use can be further examined in multilevel covariance analysis (Muthën & Muthën, 1998; Wan, 2002).

CONCLUSIONS

A structural equation model evaluates whether the proposed relationships (here, among social capital, health status and health services use) are consistent with the actual patterns found in the empirical data, through a two-step process: the measurement model and the structural equation model. The implementation of SEM analytical approach is particularly appropriate in situations where multiple constructs and indicators (e.g. health status) are taken into account at the same time.

With the aid of the user-friendly software program AMOS, it was found that among the relationships of social capital factors and the health variables of health status, and use of health services, only health status demonstrates statistical significance in the structural equation models of health services use. The results show that health status as indicated by six indicators of health from SF-12 outweighs social capital factors in predicting health services use. To some extent, this study supports the presumption that a generic psychometric scale such as SF-12 has predictive power for health services use.

The effect of social capital on use of different health services deserves further investigation. On the one hand, use is associated with access to care, and thus one cannot ignore the importance of social capital viewed at the community or ecological level in predicting differential use of specific types of health services. On the other hand, the sole use of social capital as a predictor of health care use raises the concerns about accuracy and equity noted in this study. In addition, the effect of social capital cannot be accurately estimated without examining it at both aggregate and individual levels by multilevel analysis.

ACKNOWLEDGMENTS

This study is part of a larger project, the Breast Cancer Screening Project in Kazakhstan, undertaken by the Williamson Institute for Health Studies at Virginia Commonwealth University, Medical College of Virginia Campus. Appreciation must be expressed to our collaborators and colleagues at the Kazakhstan School of Public Health – Drs. Alexander Chen, Altyn Aringazina, Saule Nukusheva, Askar Chukmaitov, Serik Chukmaitov, and Maksut Kulzhanov.

REFERENCES

Andersen, R. (1995). Revisiting the behavioral model and access to medical care: Does it matter? *Journal of Health and Social Behavior*, *36*, 1–10.

Andersen, R., & Newman, J. F. (1973). Societal and individual determinants of medical care utilization in the United States. *Milbank Memorial Fund Quarterly*, *51*, 95–124.

Arbuckle, J. L., & Wothke, W. (1997). *Amos users' guide*. Chicago, IL: SPSS.

Bentler, P. M., & Bonett, D. G. (1980). Significance tests and goodness of fit in the analysis of covariance structure. *Psychological Bulletin*, *88*, 588–606.

Berkman, L. F., & Kawachi, I. (Eds) (2000). *Social epidemiology*. London: Oxford University Press.

Bollen, K. (1989). *Structural equations with latent variables*. New York: Wiley.

Bourdieu, P. (1986). The forms of capital. In: J. G. Richard (Ed.), *The Handbook of Theory: Research for Sociology of Education*. New York: Greenwood Press.

Coleman, J. S. (1990). *Foundations of social theory*. Cambridge, MA: Harvard University Press.

Epstein, A. M., Stern, R. S., Tognetti, J., Begg, C. B., Hartley, R. M., Cumella, E., Jr., & Ayanian, J. Z. (1988). The association of patients' socioeconomic characteristics with the length of hospital stay and hospital charges within diagnosis-related groups. *New England Journal Medicine*, *318*(24), 1579–1585.

Gold, R., Kennedy, B., Connell, F., & Kawachi, I. (2002). Teen births, income inequality, and social capital: Developing an understanding of causal pathway. *Health & Place*, *8*, 7–83.

Hendryx, M. S., Ahern, M. M., Lovrich, N. P., & McCurdy, A. H. (2002). Access to health care and community social capital. *Health Services Research*, *37*(1), 87–103.

Jöreskog, K., & Sörbom, D. (1989). *LISREL 7: A Guide to the program and applications* (2nd ed.). Chicago, IL: SPSS.

Katz, S. J., Hofer, T. P., & Manning, W. G. (1996). Hospital utilization in Ontario and the United States: The impact of socioeconomic status and health status. *Canadian Journal of Public Health*, *87*, 253–256.

Kawachi, I., & Berkman, L. (2000). Social cohesion, social capital, and health. In: L. F. Berkman & I. Kawachi (Eds), *Social Epidemiology*. New York: Oxford University Press.

Kawachi, I., Kennedy, B. P., Lochner, K., & Prothro-Stith, D. (1997). Social capital. income equity, and mortality. *American Journal of Public Health*, *87*(9), 1491–1498.

Kawachi, I., Kennedy, B. P. et al. (1999). Social capital and self-rated health: A contextual analysis. *American Journal of Public Health*, *89*(8), 1187–1193.

Kennedy, B. P., Kawachi, I., Prothrow-Stith, D., Lochner, K., & Gupta, V. (1998). Social capital, income inequality, and firearm violent crime. *Social Science and Medicine*, *47*, 7–17.

Lochner, K. A., Kawachi, I., Brennan, R. T., & Buka, S. L. (2002). Social capital and neighborhood mortality rates in Chicago. *Social Science and Medicine*, *51*, 1–9.

Muthën, L. K., & Muthën, B. (1998). *Mplus user's guide: The comprehensive modeling program for applied researchers*. Los Angeles: Muthën & Muthën.

Putnam, R. D. (1993). *Making democracy work: Civic traditions in modern Italy*. Princeton, NJ: Princeton University Press.

Putnam, R. D. (1995). Bowling alone: America's declining social capital. *Journal of Democracy*, *6*(1), 35–42.

Sampsom, R. J., Raudenbush, S. W. et al. (1997). Neighborhoods and violent crime: A multilevel study of collective efficacy. *Science*, *277*(5328), 918–924.

Stump, T. E., Johnson, R. J., & Wolinsky, F. D. (1995). Changes in physician utilization over time among older adults. *Journal of Gerontology, Social Sciences*, *50B*, S45–S58.

Tucker, L. R., & Lesis, C. (1973). A reliability coefficient for maximum likelihood factor analysis. *Psychometrika*, *38*, 1–10.

Veenstra, G. (2002). Social capital and health. *Social Science & Medicine*, *54*(6), 849–868.

Wan, T. T. H. (1989). A behavioral model of health services utilization by older people. In: Marcia Ory & K. Bond. (Eds), *Aging and Health Care*. London: Routledge.

Wan, T. T. H. (2002). *Evidence-based health care management: Multivariate modeling approaches*. Boston: Kluwer Academic Publishers.

Ware, J. E., Kosinski, M., & Keller, S. D. (1995). *SF-12: How to score the SF-12 physical and mental summary scales* (2nd ed.). Boston, MA: The Health Institute, New England Medical Center.

Ware, J. E., Snow, K. K., Kosinski, M., & Gandek, B. (1993). *SF-36 health survey: Manual & interpretation guid*. Boston, MA: Nimrod Press.

Weitzman, E. R., & Kawachi, I. (2000). Giving means receiving: The protective effect of social capital on binge drinking on college campuses. *American Journal of Public Health*, *90*(12), 1936–1939.

Wolinsky, F. D., & Johnson, R. J. (1991). The use of health services by older adults. *Journal of Gerontology: Social Sciences*, *46*, S345–S357.

PART IV: BROADER POLICY CONCERNS AND HEALTH INSURANCE REFORM

UNIVERSAL HEALTH CARE – AN IDEA WHOSE TIME HAS COME

Elizabeth Furlong

ABSTRACT

Health care systems are evaluated by the triad of access, quality, and cost. This article presents evidence-based outcomes of multiple measures of concern with the United States (U.S.) health care system and proposes a universal health care system as the solution of choice. One third of the U.S. population is either non-insured or underinsured. Lack of quality care is shown by several indicators. Cost concerns are noted in cross-national studies which emphasize that the U.S. spends twice as much for health care but with less access for its citizens to health care. The presidential election of Fall 2004 provides a "window of opportunity" for this policy to be enacted.

INTRODUCTION

Universal health care is "an idea whose time has come" (Kingdon, 1995, p. 1). This chapter will articulate why this author believes universal health care is important and necessary at this time for the United States. Indicators of concern will be discussed – cross-national comparison of health indicators, measures of both clinical and financial issues in the U.S., statistics on health disparities, and, increasing concerns by many about lack of access to care. Following this, solutions will be presented as articulated by the Physicians for a National Health Program

Reorganizing Health Care Delivery Systems: Problems of Managed Care and Other Models of Health Care Delivery
Research in the Sociology of Health Care, Volume 21, 183–193

ISSN: 0275-4959/doi:10.1016/S0275-4959(03)21010-3

(PNHP) and by some 2004 U.S. presidential candidates. Analysis will be done of these positions. While research-based studies and evidence-based outcomes will be the major theme of documentation for this article, the author will intersperse some anecdotal stories to "to put a face" on the concerns noted in this paper.

A parish nurse/health minister in a large Protestant church in a mid-western city called this author and voiced concern with the current health care delivery system (Stephanie Ullrich, April 30th, 2003). While her parish is located in a middle class neighborhood, she is frustrated with her inability to successfully resolve some health situations for her parishioners. She and the church's ministers are concerned with the lack of insurance for all, with "people falling through the cracks," and with the injustices of the health system. She is interested in being a part of "something larger" and/or groups or coalitions that are working at system change. The "flashpoint event" for she and her colleague ministers is the recent closing in this city of the major hospital for mentally ill (MI) patients. She is now hearing the personal stories of the MI population being "dumped" at Francis House, a Catholic Worker House shelter for the homeless. She recognizes that the health system is not working for many individuals and groups. To understand some of these "larger" issues, data will be presented on the extent of various problems within the health system.

LACK OF ACCESS

A major problem is the number of people who are not insured. There is 17% of the population without health insurance – with another 17% who are underinsured (Woolhandler & Himmelstein, 2000, slide 1). This is a linear trend that has increased since 1975. It is important to note that of the 40–50 million people not insured, 50% are employed, 25% are children, 5% are unemployed, and 20% are not in the work force (Woolhandler & Himmelstein, 2000, slide 2). These statistics are important – especially the populations of workers who are not insured and the population of children without health insurance coverage. Employees are not insured for a variety of reasons – the employer may not offer insurance, insurance may be offered but may not be affordable by the employee, and so forth. Thus, American voters need to be aware that many other Americans, even while working, can not afford health insurance. This writer gave a presentation on universal health care to a group of Catholic deacons and their wives in a small rural community in Nebraska. While this group was not receptive to the idea of universal health care and federal government involvement in health care, it was poignant to this writer to hear the story of one wife during a break. While she shared her story privately with me, she exemplified the concern with lack of health insurance. Both she and her husband worked and her husband had health insurance through his work site. Insurance was not offered at her worksite, and, as a couple, they could not afford to purchase the additional insurance through his worksite plan. As a diabetic, she literally thanks God daily for not being hospitalized – as that would cause bankruptcy for them. She shared her story because she was interested in learning about resources where she could find health insurance at a cost affordable to them. The second statistic is also of concern, that of children, because they are a vulnerable population.

This tradition in the U.S. of health insurance being tied to one's employment status dates to World War II. The U.S. is the only First World industrialized country that has this system. Other such countries provide universal health care for their citizens. This system began because of the salary caps put on employee salaries during the war. However, benefits could be given to employees – thus, health insurance was promoted by unions at that time as such a benefit. Even though health insurance status is tied to employment status in this country, it is not consistent. As noted above, it depends on what employment setting you work in, what the specific benefits are, and so forth. Individual business owners, people working part-time, people working in small businesses, and so forth are frequently left with no health insurance. For those employees who can purchase health insurance at their worksite, the cost of health insurance premiums has increased and the cost has shifted from employers to employees (Sheils, Hogan & Manolor, 1999). For example, the percent of premiums paid by employees increased from 10% in 1988 to 20% in 1996 (Sheils, Hogan & Manolor, 1999).

HEALTH DISPARITIES

Besides the large population of people not being insured (17%) and many employed people not being insured, another problematic area is the area of health disparities. Healthy People (2010) has set a goal to eliminate health disparities (Healthy People, 2010). It is noteworthy that the goal verb is "eliminate" versus "decrease." However, the disparities in non-health insurance coverage by ethnic background is significant – 32% for Hispanics, 18% for African-Americans, and 10% for Caucasians (Woolhandler & Himmelstein, 2000, slide 4). An analysis of this data is even more dramatic when applying this to the changing demographics of this country, i.e. the rapid increase in the Hispanic population. The correlation of this health disparity with clinical evidence-based outcomes will be noted in a later section of this paper.

A face was put on the concerns for Hispanic immigrants with no insurance by a newspaper report of a major new convention center being constructed in a mid-western city (Jordan & Gonzalez, 2003). Contractors and sub-contractors employ illegal immigrants for such positions. While there are advantages for the workers, the subcontractors, the general contractors, the taxpayers, and immigrants' families in Central and South America, there are also limitations. For example, health providers and health organizations who care for such individuals either pass the costs on to the insured population or may suffer economically themselves. Or worse, the individuals receive no care nor Workman's Compensation. The newspaper journalist reported the general contractors denied responsibility for the sub-contractors' choice of employing illegal immigrants. In effect, U.S. taxpayers are subsidizing the subcontractors.

INCREASED COSTS

Thus far, this paper has addressed several populations and their concerns: (A) the one third of Americans without health insurance (17% without health insurance and 17% who are underinsured); and (B) ethnic minority populations who are experiencing health disparities. In part, this is due to their not having health insurance, which then limits their access to health care. Another sub-set of the U.S. population with concerns is the older population, i.e. those on Medicare and/or Medicaid. They, like employees paying more for health insurance premiums, are also paying more for out-of-pocket health expenses (Woolhandler & Himmelstein, 2000, slide 7). The percent of income that seniors pay for health care has doubled from 1977 to 2000 from 12 to 25% (Woolhandler & Himmelstein, 2000, slide 7). For this population, a great concern is how long-term care will be paid for. Many Americans are not knowledgeable in this area and mistakenly believe that Medicare will pay for such care. The two largest funders for long-term care are Medicaid which pays for 44% of long-term care and private pay which re-imburses 38% of such care (Feder, Komisar & Niefield, 2000). Medicare will only pay for long-term care if there is "skilled" care being given by one of these professionals – nursing, speech, physical therapy, or occupational therapy. Some research has shown that the average time it takes for Americans to "spend down" their resources of private pay is six months at which time they become eligible for Medicaid.

Another concern of the older population is the increasing cost of prescription drugs. U.S. drug spending has doubled in an eight year time span from 1990 to 1998 (Woolhandler & Himmelstein, 2000, slide 84). Pharmacists predict that drug use and costs will keep increasing because of many factors – new knowledge, the emerging "Baby Boomer" population, direct television advertising to consumers, and so forth. The lay, health, and policy literature has been replete the last several years about the debate on this concern at the Congressional level. To date, this one aspect of the health care system has not been resolved.

MEDICAID CRISIS

Perhaps the population hurting the most in 2003 is the group who depends on Medicaid for health insurance. For people receiving health care through Medicaid, the situation is especially bleak during the state fiscal crises being experienced by all states. "The issue is already roiling state legislatures . . . Many state officials are pleading for federal help as they face an array of painful trade-offs, often pitting the needs of impoverished elderly people for prescription drugs and long-term case against those of low-income families seeking basic health coverage" (Toner & Pear, 2003, p. 1). The following is a compilation of many Medicaid changes recently made or planned by many states: (1) 1.7 million Americans may lose coverage; (2) dental and vision coverage may be eliminated in California, Florida, and Ohio; (3) Mississippi and Oklahoma will decrease the number of prescriptions paid for; (4) Kentucky and Massachusetts will tighten criteria for eligibility for long-term care; (5) 240,000 children and 17,000 pregnant women in Texas may lose coverage; (6) Tennessee has removed 55,000 children and a total of 208,000 people from its Medicaid program known as TennCare; (7) Oklahoma eliminated its "medically needy" program for 8,300 people with catastrophic medical expenses; (8) Colorado terminated Medicaid benefits for 3,500 legal immigrants, and so forth (Toner & Pear, 2003).

Another population who suffers in the current system is those without health insurance who have to file bankruptcy because of medical bills. The lack of health insurance leads to monetary stresses for families. One indicator of this is the research on bankruptcies. From a legal financial perspective, the majority of bankruptcies are due to medical illnesses (Woolhandler & Himmelstein, 2000, slide 9). Forty-five percent of all bankruptcies are due to this.

NEGATIVE CLINICAL OUTCOMES

The health care system is being increasingly driven by the phenomenon of evidence-based outcomes, i.e. medical interventions are selected based on research-based outcomes. Further, any health care system is evaluated in terms of the triad measures of access, quality, and cost. The earlier part of this paper has addressed the lack of access. There is also much research which correlates the status of being insured with clinical outcomes. For example, there is a correlation with having insurance and having access to health care. Almost twice as many people with health insurance versus those without health insurance receive care for serious and potentially serious symptoms (Woolhandler & Himmelstein, 2000, slide 12). A second example is research on the population of pregnant women. The major factor for not seeking prenatal care is having no insurance nor money (Woolhandler & Himmelstein, 2000, slide 15). Thus, the system problem of lack of access leads to negative indicators of poor quality of care as measured by outcome data.

Some populations do not have health insurance while other populations vary in their health insurance programs and options. For example, there is a correlation between the amount of choice that one has in insurance options and one's salary. This is a linear trend – the higher the salary, the more choices one has (Gawande, Blendon, Hugick, Brodie, Benson & Levitt, 1998). This particular research needs to be known by Americans. Americans value choice and may believe they have choice with health insurance plans. However, the research does not reflect this. This research is important because of the political power of the Harry and Louise ads in the early 1990s which derailed the Clinton Plan for a revised health care system. The use of those media ads was very successful in influencing Americans that they would lose choice if they supported that plan. However, the research shows that Americans do not have the degree of choice that they perceive. Critical thinking, analysis, and education on this aspect needs to be done.

SYSTEM FINANCIAL CONCERNS

Besides the above access and quality concerns, there are also systematic cost concerns. The current system is a for-profit system. The question needs to be asked – is it ethical to make a profit off of the suffering of others? Salaries and stock options for CEOs of health insurance and managed care organizations range from salaries of 0.025 million dollars to five million dollars and stock option values of 445 million dollars to 64 million dollars (Woolhandler & Himmelstein, 2000, slide 62). In addition, the amount of overhead and profit spent by these organizations range from 14 to 33% (Woolhandler & Himmelstein, 2000, slide 63). Such costs stand in contrast to Medicare which spends 5% for administrative overhead. The PHNP evaluates that much of the cost of converting to universal health care can be captured by the savings in the current overhead and profit sectors.

CROSS-NATIONAL COMPARISONS

Cross-national comparisons of the U.S. with other First World industrialized democracies offers other compelling data. The U.S. is the only such country that does not provide universal health coverage for its population. From a public health perspective, one uses the health indicators of infant mortality rate and maternal mortality rate to evaluate the progress of a country because these two measure a country's progress and the value it places on vulnerable populations. When ranked with like countries, the U.S. rates the highest in infant mortality rate (Woolhandler & Himmelstein, 2000, slide 92). At the opposite end of the life continuum, again, the U.S. rates poorly. Life expectancy is the lowest for both men and women when compared to like countries (Woolhandler & Himmelstein, 2000, slides 93 and 94). In addition, there are two other indicators of concern. Survey data reveals that when asked about the difficulty in obtaining care, individuals find it the most difficult in the U.S. (Woolhandler & Himmelstein, 2000, slide 103). A final indicator is survey data on people's experience with continuity of care with a primary care provider. Individuals in the U.S. have the least continuity of care (Woolhandler & Himmelstein, 2000, slide 104). This is another most important finding because of the extensive amount of research on the outcomes of having continuity of care with a primary care provider – especially for those individuals with a chronic disease and for children. The studies are consistent – there is less morbidity and mortality when one has continuity of care with a primary care provider.

Another cross-national comparison is the systemic financial aspect. In 2001, health care costs were 14% of the gross national product. It is predicted to rise to 18% by 2012. While other like countries assure insurance coverage for all their citizens which the U.S. does not, the cost of spending on the health care system is the highest for the U.S. (Anderson, Hurst, Hussey & Jee-Hughes, 2000). Put another way, the U.S. spends about double the cost per capita – but, with worse outcomes.

A recent report from the Institute of Medicine has analyzed the impact of being uninsured on the larger community (Institute of Medicine, 2003). They found that hospitals in urban areas with high uninsured rates have less total inpatient capacity, fewer services for vulnerable populations, and are less likely to offer trauma and burn care. And, conversely, in rural areas with high numbers of uninsured people, there are lower financial margins, fewer intensive care beds, fewer high-technology services, and fewer inpatient psychiatric services. In both cases, they note the spillover into the local economy, the spread of communicable disease, shortage of healthcare providers and reduction in public health spending.

There are many other concerns that could be identified with the U.S. health care system, i.e. health care being denied to some by managed care organizations, the proliferation of state legislation since the demise of the Clinton Plan to address abuses with managed care organizations, the sub-set of the population who are mentally ill who are not served well, erosion of the physician/patient relationship, and so forth (Himmelstein, Woolhandler, Hellander & Wolfe, 1999). In summary, these are the concerns articulated thus far – one third of the U.S. population are either uninsured or underinsured, there are clinical indicators of concerns, and the U.S. fares poorly in clinical and financial outcomes when compared to other like countries.

UNIVERSAL HEALTH CARE AS A SOLUTION

Before promoting universal health care as the answer to the current crisis in the health care system, the following synthesis will be done of the many problems identified above. Incremental change has not worked. Solving one part of the health care system frequently only results in new problems in another part of the system. The many changes occurring in the Medicaid system will be used as an exemplar. People who lose Medicaid insurance may rely on the emergency department because they do not have insurance nor money to seek care in a more appropriate manner. The hospital either loses money or passes its costs on to other patients. Thus, other Americans are subsidizing the cost. Or if a person is covered by Medicaid but Medicaid decreases payments to health providers, those primary care providers may choose not to take such patients. Thus, those Medicaid patients have no primary care provider from whom to seek care. Their morbidity and mortality increases because of not having continuity with a primary health care provider. Further, they may seek care from an emergency room because of lack of care elsewhere. In Indiana, physicians have not had a rate increase since 1994. (Toner & Pear, 2003). Medicaid has been decreasing payment rates to nursing homes; in turn, they have reduced staff, decreased staff education, eliminated liability insurance, etc. (Toner & Pear, 2003). These few examples – with just the 2003 Medicaid crisis demonstrate that changes and ripples in one sector of the health care system – cause many other changes to occur. Problem solving has to be systemic.

This author argues the answer to the many problems identified above is a universal health care system. One solution is that promoted by the PHNP, a group of 10,000 physicians. They acknowledge that Americans are interested in these traits for a health care system – guaranteed access, choice of physicians, high quality, affordability, trust and respect (Woolhandler & Himmelstein, 2000, slide 124). Their analysis is such a system can be best accomplished by a universal health care system. Their model would be characterized by the following: (1) it would be universal and cover everyone; (2) there would be no co-pays; (3) comprehensive care would be provided; (4) administration would be a single public payer; (5) the profit would be removed from the system; (6) improved health planning; (7) public accountability for quality and cost; (8) minimal bureaucracy; (9) choice of providers; (10) physicians and health organizations would negotiate fees and budgets; (11) local planning Boards; (12) progressive taxes going to a Health Care Trust fund; and (13) a public agency to process and pay the bills (Woolhandler & Himmelstein, 2000, slides 125 and 126).

While the PHNP argues it is possible because it is known that universal health care is being implemented in other like countries, analysis of the barriers also needs to be done. A difference between the U.S. and Western European countries is one's attitude toward government, and, especially toward the federal government. Americans dislike governmental intervention. On the other hand, Europeans have a positive attitude toward government. This difference in ideology has had important implications for Americans' acceptance or rejection of governmental involvement in their daily lives. "And, a political culture deeply distrustful of government raises a high barrier to the kind of centralized health care systems found in most other advanced industrial countries" (Broder, *Lincoln Journal*, April 3rd, 2003). This particular barrier needs to be addressed. Trend lines that argue in favor of a changing mentality among Americans about universal health care are these: (1) a 1998 Harris poll that reflected three fourths of Americans agreed "the government should provide quality medical coverage to all adults"; (2) a 1999 poll of medical students and faculty reflected that 57% of that population favored a single payer system; and (3) the demise of the dot.com economy and an increasing number of middle class educated workers losing jobs in the past two years has "put a new face" on the individuals without health insurance, and so forth (Simon, Pan, Sullivan, Clark-Chiarelly, Connelly, Petus, Singer, Inui & Black, 1999; Woolhandler & Himmelstein, 2000, slide 130).

Policy activists are analyzing not only the PHNP position but also some of the U.S. presidential candidates' health care positions. In recent months both former Vermont governor, Howard Dean, MD and Congressman Richard Gephardt have emphasized their health care plans. Governor Dean has a track record of expanding health care coverage to almost all children in that state. Both candidates would use the current private sector and are not espousing the plan by PHNP.

Researchers have analyzed another solution – that of President Bush's tax credit proposal which facilitates access for health insurance through a policy of tax credits. However, a critical economic analysis of this plan demonstrates that it does not work for families – i.e. they would not be able to afford health insurance (Personal communication, Richard O'Brien, March 12th, 2003).

Some Congressmen have introduced legislation for a single payer system. That was introduced by Congressmen Conyers (D-MI) and Kucinich (D-OH) in February 2003 (Personal correspondence, Quentin Young, January 2003).

As the debate continues on how to best provide health care for the U.S. population, conversation will focus on the three corners of the health triad – cost, quality, and access. This paper has argued that the current U.S. system is not measuring well when evaluating all three indicators. The U.S. spends twice as much as other like countries with worse clinical outcomes and with one third of its population either not insured or underinsured. One can also expect to hear discussion on the politics and ethics of the current and proposed systems. Benatar argues that "bioethicists may have become co-opted . . . and critical attitudes to the commercialization of health have been muted" (2003). Further, his writing, from this author's perspective, is supportive of the PHNP – "the legitimacy of the USA's heavily market-oriented health care system, which accounts for 50% of total annual global health care expenditure on 5% of the world's population and yet excludes many of its citizens (while claiming to be a standard to which others should aspire), needs to be questioned and contested" (Benatar, 2001).

SUMMARY

In summary, given the problems of the U.S. health care system, universal health care is "an idea whose time has come." There is a timely "window of opportunity" because of the upcoming Fall 2004 presidential elections where health care issues are being increasingly discussed by President Bush and the Democratic presidential candidates. On an ongoing basis Congressmen are also discussing health concerns and proposing policy. The beginning years of the first decade of the 21st century are different from the early years of the 1990s when the Clinton health plan was introduced. Besides all of the documented evidence in this paper which points to a need for change, other indicators of a soft economy, the loss of jobs by many middle class individuals (with loss of health insurance) and increasing costs for both employers and employees may propel the American public to be more open to governmental intervention and a universal health care plan. Further, all of the ramifications since September 11th may be reasons why Americans could recognize that there is need for governmental interventions in certain sectors of society. This author argues that a universal health care plan, such as proposed by PHNP, is the best solution for this country – to best meet cost, quality, and access indicators of health care. Incremental health policy change in a mixed private/public health system has not worked. It is time to implement the policy of universal health care and for the U.S. population to have and to enjoy the kind of total population health benefits enjoyed by people in other First World industrialized countries.

REFERENCES

Anderson, G. F., Hurst, J., Hussey, P. S., & Jee-Hughes, M. (2000). Health spending and outcomes: trends in OECD countries, 1960–1998. *Health Affairs*, *19*(3), 150–157.

Broder, T. (2003, April 3rd). *Lincoln Journal.*

Benatar, S. R. (2001). Promoting national and international justice through bioethics. *BMC Medical Ethics, 2,* 2.

Feder, J., Komisar, H. L., & Niefield, M. (2000). Longterm cae in the U.S.: An overview. *Health Affairs, 19*(3), 40–56.

Gawande, A. A., Blendon, R., Hugik, L., Brodie, M., Benson, J. M., & Levitt, L. (1998). Does dissatisfaction with health plans stem from having no choices? *Health Affairs, 17*(5), 184–194.

Healthy People (2010). U.S. Government.

Himmelstein, D. U., Woolhandler, S., Hellander, I., & Wolfe, S. M. (1999). Quality of care in investor-owned vs. not-for-profit HMOs. *Journal of American Medical Association, 282*(2), 159–163.

Institute of Medicine Report. Retrieved April 16, 2003, from http://www4.nationalacademies.org/oni/webextra.nsf/web/uninsured

Jordan, S., & Gonzalez, C. (2003, April 27th). When a worker is not an employee. *Omaha World Herald.*

Kingdon (1995). *Agendas, alternatives, and public policies.* New York: Harper, Collins College Publisher.

Sheils, J. F., Hogan, P., & Manolov, N. (1999). Paying more and losing ground: How employer cost-shifting is eroding health coverage of working families. *International Journal of Health Services, 29*(3), 485–518.

Simon, S. R., Pan, R. J., Sullivan, A. M., Clark-Chiarelly, N., Connelly, M. T., Petus, A. S., Singer, J. D., Inui, T. S., & Black, S. D. (1999). View of managed care – a survey of students, residents, faculty, and deans at medical schools in the U.S. *New England Journal of Medicine, 340,* 928–936.

Toner, R., & Pear, R. (2003, April 28th). Cutbacks imperil health coverage for states' poor. *New York Times,* p. 1.

Woolhandler, S., & Himmelstein, D. U. (2000). *The national health program slideshow guide.* Cambridge, MA: Center for National Health Program Studies, Harvard Medical School.

LIFE AFTER WELFARE IN RURAL COMMUNITIES AND SMALL TOWNS: PLANNING FOR HEALTH INSURANCE

Karen Seccombe and Richard Lockwood

ABSTRACT

This research explores how families coming off of Temporary Assistance to Needy Families (TANF), the national cash welfare program, plan for their health insurance after their automatic benefits expire. Data were collected in focus groups in rural communities and small towns in Oregon. Respondents reported that topics related to health insurance or planning for health insurance are not components of any welfare-to work curriculum, nor are they part of routine conversations with caseworkers. Many respondents reported that we were the first ones to raise these issues with them. Consequently, they had done virtually no planning for when their transitional Medicaid expires despite their serious concerns about access to health care and their previous negative experiences with being uninsured.

INTRODUCTION

There is a push to get people off of welfare. Welfare programs are unpopular, and have been accused of fostering long-term dependency, family breakups, and illegitimacy (Murray, 1984, 1988). Former President Clinton signed sweeping

Reorganizing Health Care Delivery Systems: Problems of Managed Care and Other Models of Health Care Delivery
Research in the Sociology of Health Care, Volume 21, 195–210

ISSN: 0275-4959/doi:10.1016/S0275-4959(03)21011-5

welfare reform legislation, which became Federal law on July 1st, 1997. Turning many of the details of welfare law over to states, Temporary Assistance to Needy Families (TANF) set lifetime welfare payments at a maximum of five years, with the majority of adult recipients being required to work after two years. Many states have adopted significantly shorter time frames, and some make no allowances for women who have young babies. Moreover, twenty states have also adopted some form of family caps, meaning that benefits will not be increased for a child born while a mother is already receiving TANF (Rowe, 2000).

Since the passage of welfare reform, many people have left welfare, usually for low-wage work. These changes, coupled with the expanding economy and low unemployment rate of the 1990s, have contributed to a dramatic initial drop in the number of families receiving TANF. From March 1994 to June 2002, national caseloads fell by 60%, declining from 5 to 2 million families (U.S. Department of Health and Human Services, 2002).

How do families intend to secure their health insurance after leaving welfare? Families continue to qualify for one year of automatic transitional Medicaid coverage, but after that they may lose the health insurance that was a critical part of their welfare benefits. This research is based on data gleaned from focus groups with welfare recipients in welfare-to-work programs, and examines how adults leaving welfare plan for the termination of their Medicaid benefits.

Background: Welfare Reform and Health

Early data indicate that the recent drop in welfare caseloads has been accompanied by a decrease in the number of people receiving Medicaid (Greenberg, 1998; Larkin, 1999). Despite the creation of SCHIP (State Children's Health Insurance Program) which is designed to extend health insurance coverage for children up to 200% of poverty and enrolls about 3.5 million low-income children, the majority of people who lost coverage were children under 19 (Families USA, 1999; Klein, 2002). Approximately one-half of children lose Medicaid coverage when their families leave welfare. Two million children who are eligible for coverage through the SCHIP program remain uninsured. The reasons for the drop in Medicaid and SCHIP coverage are unclear, but it could be due, in part, to lack of knowledge, confusion, or error on the part of either recipients or their caseworkers regarding eligibility. Moreover, a combination of factors – pending reductions in federal funding, the reversion of previously allotted funds back to the U.S. Treasury, and the growing state budget crises – could reduce program enrollment and increase the number of uninsured children. The Bush administration conservatively estimates that SCHIP will drop by 900,000 between 2003 and 2006 (Klein, 2002).

This decline in Medicaid and SCHIP coverage would be of less concern if we could assume that former welfare recipients and their children receive health insurance from their new employers. However, this is generally not the case. The low-wage jobs that most welfare recipients take often do not provide health insurance as a fringe benefit (Loprest, 2001, 2002; O'Brien & Feder, 1998). Moreover, the erosion of employer-sponsored health insurance coverage has occurred most dramatically among workers with the lowest wages (Seccombe, 1995). Given that most former welfare recipients are employed in jobs that pay approximately $7.15 an hour according to a national survey, it is unlikely that these families have the resources to purchase health insurance privately (Loprest, 2002). Medicaid is not an option for many: A parent in a family of three working full time at $7 an hour would not qualify for Medicaid in 28 states. In 13 other states, the maximum hours a parent can work at $7 an hour and still qualify is less than 20 hours per week (Kaiser Commission on Medicaid and the Uninsured, July, 2002).

Being without insurance is problematic for poor families. Poor adults and children are more likely to suffer a wide variety of serious chronic and acute ailments than are their non-poor counterparts (Families USA, 2001; Heymann & Earle, 1999; Kaiser Commission on Medicaid and the Uninsured, July 2002; Loprest & Acs, 1996). For example, poor children are more likely to be iron deficient, to have frequent diarrhea or colitis, to have asthma and lead paint poisoning, to suffer from partial or complete blindness or deafness, and are more likely to die in childhood. Without insurance, both adults and children are less likely to have a regular source of health care, are more likely to rely upon emergency rooms for their treatment, and often experience unnecessary pain, suffering, and even death (Kaiser Commission on Medicaid and the Uninsured, 1998; Weigers, Weinick & Cohen, 1998). The results of a Kaiser/Commonwealth 1997 National Survey of Health Insurance indicate that 55% of people without insurance postponed getting medical care, compared to only 14% of people with insurance. Thirty percent of the uninsured reported that they did not get needed medical care, compared to only 7% with insurance. And 24% reported that they did not fill a prescription, compared to 6% of people with insurance. Not surprisingly then, welfare recipients report that Medicaid is their most important benefit – more important than food stamps, subsidized housing, or the welfare check itself (Seccombe, 1999).

The welfare reform bill allows for 12 months of transitional Medicaid assistance for former welfare recipients and their families who would otherwise lose eligibility because of their earnings. But studies reveal that many families coming off of welfare do not have health insurance after their one year of transitional Medicaid expires (Families USA, 1999; Garrett & Holahan, 2000; Loprest, 2001, 2002). They report that at least one-quarter to one-third of adults, and approximately one-quarter of children are completely uninsured after their transitional Medicaid expires. More importantly, no one really knows whether families are getting the health care that they need. In one study conducted by the state of Washington, uninsured families report that they are worse off now with respect to health and well-being since leaving TANF, whereas families who have health insurance do not.

This research addresses the question of how families leaving welfare plan for the termination of their Medicaid benefits. In particular, we focus our inquiry in the state of Oregon. Oregon is a unique state because it is the site of one of the more progressive state-funded health care programs in the nation, known as the Oregon Health Plan (OHP). It is considered a leader by other states and carefully observed. Thus, the results from this study have important implications for other states as they consider ways of providing former welfare recipients and their children with secure access to health care.

The State of Oregon: Site of the Oregon Health Plan

The OHP is designed to offer limited health insurance coverage to low-income adults and children at little or no cost, including those who leave welfare. It was developed to assist the working poor in accessing the health care system by offering them insurance that would cover common medical procedures. Benefits include medical, dental, and chemical dependency services. Income guidelines are such that a family of four would need gross monthly incomes of $1,421 or less in order for all family members to qualify; $1,890 for coverage for children under six and pregnant women; and $2,415 for coverage for pregnant women and their unborn child. The OHP has been touted as providing access to health care for 500,000 people, including more than 15,000 individuals who had previously been denied coverage due to pre-existing medical conditions. The percentage of people in Oregon without insurance dropped from 18% in 1990 to 11% in 1998, while it remains 16% nationally. Likewise, the percentage of uninsured children in Oregon dropped from 21% in 1990, to 12% in 2000, compared to 16% nationally (Oregon Health Plan Policy Research, 2003). Not surprisingly then, other states watch the OHP carefully, hoping that it will provide a model for significantly reducing the proportion of uninsured within the population, and opening up access to the health care system.

Consequently, because of the OHP it is less clear how welfare reform trends will affect the size of the uninsured population. Theoretically at least, former recipients should be able to buy into the OHP if they work at jobs that do not provide health insurance and have relatively low earnings. But apparently, some people do not choose to do this, are not eligible for assistance, or are unfamiliar with their entitlements, given the fact that thousands of people in Oregon are still uninsured.

Vulnerable Families: Living in Rural Communities and Small Towns

A recent survey found that the problem is particularly acute among rural residents, including children. For example, while 4% of children in the Portland metropolitan area are without insurance, approximately 16% of children are uninsured in the rural eastern regions of the state. In rural communities and small towns, employers are less likely to offer insurance and options are more limited (Duncan, Seccombe & Amey, 1995). This is a disturbing social problem, and one that is likely to get worse as the economy falters in the 2000s and state budgets decline.

How do these vulnerable families plan for the time they leave welfare and their automatic transitional benefits will expire? What are their perceived employment prospects, and their experiences with the OHP and with being uninsured in the past? We want to hear directly from poor families in rural communities and small towns themselves rather than relying upon media stereotypes so that we can learn of their real needs, the concerns that they express, and the constraints under which they operate. Understanding how rural families plan for their insurance coverage and how they hope to meet the health care needs of their families after the termination of their automatic benefits while living in a geographic location with more limited options are important public policy concerns.

METHODS

We sought the stories of the most vulnerable people leaving welfare for which these issues are relevant. Data were collected in the summer of 2000 in seven focus groups with women and men who were transitioning off of TANF in rural regions and small communities across Oregon ($N = 35$). Focus groups were used because they are particularly advantageous in gathering data on more subtle or complex behaviors and attitudes, especially when people do not have easily accessible ways of talking about a topic (Morgan, 1997). Focus groups permit the emergence into discourse of areas of experience unanticipated by researchers. Our goals were to provide respondents with the maximum opportunity to shape our research from the ground up.

Segmentation

The primary segmentation criterion was geographic region. Rural regions were selected because of their residents' greater than average likelihood of being uninsured, and the more limited job opportunities for families leaving welfare for work. The specific communities were theoretically selected because of their

Table 1. Characteristics of Communities in Which Focus Groups were Conducted.

	Albany	Astoria	Hood River	Ontario	The Dalles
Population	40,010	9,990	5,135	10,910	11,880
Percentage Hispanic	4%	3%	24%	29%	8%
Percentage Black[a]	<1%	<1%	1%	<1%	<1%
Poverty Rate (County)	12.3%	13.3%	13.0%	19.6%	12.9%
Children in Poverty (County)	14.8%	17.2%	17.1%	23.4%	16.5%
Unemployment Rate (County)	8.8%	6.0%	9.6%	8.6%	7.8%
Distance to Portland Metro area (Miles)	81	100	60	376	84

[a] In no county outside of Multnomah County, site of the City of Portland, does the percentage of the population which is Black exceed 1%.

diversity in terms of race/ethnicity, poverty rates, unemployment, major industries and employers, and degree of access to a metropolitan area. However, ethnicity was not an explicit segmentation criterion for this study. The demographic profiles of the communities are reported in Table 1. The poverty rate ranged from 13 to 20% within these communities, with the child poverty rate up to 23% in one region. Unemployment ranged from 6 to 10%. The county-wide Hispanic population ranged from 3 to 29%, and the Black population constituted approximately 1% (a proportion that is representative of rural areas in Oregon).

Recruitment

Agency staff members recruited subjects from welfare-to-work programs for our focus groups. Participants were informed verbally and in writing that the agency was not involved in the project, that no specific comments would be tied to any one individual or any one focus group, that a university rather than the welfare office was conducting the research, and that future benefits were not tied to participation in the focus group. All Internal Review Board procedures were followed. The focus groups were conducted at the site of the welfare-to-work program in the area for the convenience of participants; however, staff were not present when the groups were conducted. Participants were paid $25 for their participation.

The focus groups lasted approximately 60–90 minutes. One member of the research team facilitated the focus group discussion, while another one wrote extensive notes. The focus groups were not audiotaped or videotaped.

The demographic profiles of focus group participants are found in Table 2. The ages of individual participants ranged from 17 to 47 years, with a mean of 31.5. Approximately one-third of participants were racial/ethnic minorities, primarily

Table 2. Demographic Characteristics of Focus Group Participants ($N = 35$).

Age	
Under 20	1
20–29	15
30–39	13
40–49	6
Marital status	
Never Married	12
Divorced	11
Separated	8
Married	4
Number of children	
None	1
One	4
Two	13
Three	9
Four	3
Five or more	5
Race/ethnicity	
White	23
Hispanic	9
Black	2
Native American	1
Health status	
Excellent	3
Good	17
Fair	10
Poor	5
Length of time on welfare (total)	
Less than 6 months	5
6–11 months	4
1–2 years	6
3–4 years	2
5–6 years	5
Over 6 years	10
Unknown	1
"Not on welfare"	2
Length of time on welfare (this spell)	
Less than 6 months	11
6–11 months	6
1–2 years	7
3–4 years	1
5–6 years	0
Over 6 years	3
Unknown	1
"Not on welfare"	3
Missing	3

Hispanic, as we focused on communities and regions of the state that had higher-than-average rates of ethnic minority concentration. All participants had adequate proficiency with English. Nearly one-third of participants had been on welfare during their current spell for less than 6 months, although most had spent a considerably longer period on welfare when all spells are taken into account.

Research Questions

Respondents were asked the following questions:

(1) Tell me a bit about your families' health. Do you and your children have good health, or do any of you have specific health problems?
(2) How has the Oregon Health Plan worked for you in the past? Are you able to go to the doctor when you feel you need to? Are you satisfied with the care you receive? (If no) Why not?
(3) Have you ever had a time in your life where you or your children didn't have any insurance? Were you able to get the care you needed? Did you ever postpone getting care because you could not afford it?
(4) After you go to work, where do you think you will get your health insurance for you and your family? From an employer, from the Oregon Health Plan? Did you know that health insurance from the Oregon Health Plan is not automatic after a year? Has your caseworker or case manager talked with you about this?
(5) What kind of job do you think you'll get when you go to work? How much do you think you'll earn? Do you think the job will provide health insurance for you and your family? How much do you think insurance will cost?
(6) You are all enrolled in a welfare-to-work program. How is this program helping you plan your future? Is it helping you plan for your future health care needs? What could it do to help you more?

RESULTS

Background: Health Status of Families Leaving Welfare

Paralleling national data, our focus groups reveal that families leaving welfare suffer from a variety of physical ailments, and thus the need for health care services is high. Nearly one-half of focus group participants rate their health as fair (29%) or poor (13%). Only 9% report excellent health. Chronic conditions among both adults and children are commonplace. These would require repeated health services and continued care. In particular, many respondents reported that they or their children suffer from asthma. In five out of seven focus groups, at least one incident of asthma was reported, and usually several respondents in each of these groups told us of asthma in their family. A number of other conditions were commonly reported, including allergies, ear infections, back injuries and other musculoskeletal disorders, high blood pressure, and dental problems. Other conditions that were brought to our attention included birth defects, diabetes, ulcers, lupus, immune deficiencies, heart conditions, hypoglycemia, limited vision, and addictions. These findings reflect trends in national data which indicate a greater than average need for health care services among families leaving welfare.

The Oregon Health Plan: How it is Perceived

Thirty-three of the 35 participants were covered by the Oregon Health Plan. When asked about their satisfaction with OHP, their opinions were mixed. The primary satisfaction was with knowing that there is a base floor of health care services for themselves and their families. Most respondents had spent at least some time as an adult without insurance, experiencing deleterious consequences. They had lived in other states without health insurance plans, had lived in Oregon prior to the implementation of OHP, were ineligible for OHP despite having no other coverage, or were unaware of their eligibility for OHP. The participants in our focus groups (with the exception of one woman who used alternative health care practitioners which are not covered by OHP), generally acknowledged that their lives are more secure because of OHP and that they would be worse off without it.

OHP is viewed as a valuable safety net. It gives immeasurable security to the lives of those who receive its benefits. They realized that they would likely go without coverage were it not for the program. Given the chronic nature of many of their conditions, the OHP is the only structural feature that prevents the exacerbation of their health problems. But the OHP also has its limitations. Respondents recognized a two-tiered system of health care, with OHP occupying the lower tier. OHP participants recognize that the system as currently structured does not provide any incentive to dental and health care providers to match the quality of care provided to patients covered under private insurance. This difference in incentives manifests as stigma, barriers to access, and a primary source of debt. We also saw that OHP, the hospital, and doctors' offices are conceptually indistinct in the minds of our respondents. There is a hint of a political attitude that OHP should be able to assure access to services; that is, have political influence over hospitals and doctors.

Consequences of Being Uninsured

Focus group participants were asked if they had ever gone without insurance coverage as an adult. Virtually all claimed that they had, and reported that they or their children suffered severe consequences because of it. Like the results of national studies, these consequences included such things as going without needed medical or dental care, or forgoing prescribed medicines. "Kate" summarizes the sentiment of many others when she said, "If I didn't have insurance, I didn't get help. They want their money upfront. You just suffer." Often they experienced unnecessary pain and suffering as a result. For example, one woman said that her son had dislocated his arm, but because they had no insurance and no money to pay a doctor upfront, she drove her son to a neighboring state to have her mother set it. Another woman explained that she was unable to receive necessary prenatal care, "When I was pregnant I didn't have insurance until two weeks before the baby was born. So I missed all the prenatal check-ups. It was terrible." Another woman spoke of being unable to buy needed prescriptions, "I've had some really good jobs. But some didn't have health insurance with it. There was a time I had a job and didn't qualify for OHP but didn't make enough to buy my pills. My medicine is like $8 a bottle. I couldn't buy it."

Many women spoke of the hardships that being uninsured caused for their children in particular. One described the pain her daughters experienced because of a urinary tract infection. Because they had no insurance, they postponed treatment until her daughter's pain was unbearable. "You just can't imagine the pain of hearing your daughter scream like that. Finally I had to take her to the emergency room. It's taking three months to pay the bill. I even sold my ring." She lifted her hand to show the absence of her wedding ring.

A common concern expressed in all focus groups was the cost of medical care. Most women who finally did seek treatment were unable to pay the bills. They have bad credit as a result. They accumulated a large debt, and many spoke of currently being followed by collection agencies. "My credit is bad purely because of medical bills . . ." When asked what they did about these debts, some women told me that they were making token payments, such as $20 per month, although some clinics would not accept any payment under $50. Many sheepishly told us that they were simply not paying their debts, "Can't get no blood from a turnip." Others were more indignant, such as one woman who said, "I have thousands of dollars in medical bills. Thousands. When I had my son there were problems. I had to have two epidurals and stuff. There is no way I'm paying. It's like $10,000. There is no way."

These focus groups suggest that being without insurance was a common experience before participating in the Oregon Health plan, and was a significant concern. Participants and their families experienced many deleterious consequences as a result of being uninsured, including pain, suffering, and heavy debt. The lack of insurance was seen as a significant problem and one they hoped to avoid.

Employment Prospects and Insurance

What are the employment prospects in rural communities? What is the likelihood that focus group participants will be able to secure health insurance from an employer when they leave welfare for work and after their automatic transitional benefits expire?

Fast food, hotel maids, temp agencies, clerical, childcare worker, cook, data entry, receptionist, nurses' aide – these were the types of jobs that participants hoped to find. "Good jobs" were seen as those that paid more than minimum wage by at least several dollars ($6.50 in Oregon in 2003), especially if jobs had benefits attached to them. But in small communities and rural regions, families leaving welfare face an uncertain economy with restricted job opportunities. Unemployment is high, and in several communities the work is seasonal, related to tourism or agricultural work. "I've gone all the way to Cannon Beach and Seaside to every motel applying for a maid job. No one has called me." Another woman, who has both a high school diploma and training as a nurses' aide also reports to have trouble finding work. "I applied all over this place, and then I went to (temporary agency) and they got me a job for two months. I have a CNA and a high school diploma, and I competed with lots of women for the job. A good job is $12 an hour with benefits and 37 hours a week. I applied for over 50 jobs, down to $7–$8 an hour."

Given the limited employment opportunities in their communities, they did not assume that high pay scales or health insurance benefits would be part of any employment package. Some women claimed to be searching for jobs with benefits, but they knew that in all likelihood they would not find one. This is in contrast to other focus groups conducted by the senior author in the Portland metropolitan area where optimism was high. Those participants in the metropolitan area, unlike in the rural communities we visited, felt that they would have an easy time finding work paying upwards of $10 to $15 an hour with benefits attached, enough to sustain themselves and their families. The different economic structure in urban and rural areas provided families coming off of welfare with significantly different sets of opportunities.

There was a common tendency among our focus group participants to avoid fast food work, even though the jobs are relatively plentiful. Many had worked in the industry in the past and found it demeaning working side-by-side high school students, not to mention the low pay. "I'm too old to work those jobs. . . ." They hoped for something "better." Yet participants worried that even the jobs they were likely to find – ones that paid $7 or $8 an hour – may not be sufficient to pay for their families' needs. Furthermore, many jobs start on a part-time or temporary basis and thus avoid any benefit packages. Temporary agencies were seen as a ready and good source of employment. They tend to provide higher wages – often a dollar or two above minimum wage, which was seen as a huge increase. But participants acknowledged that temp agencies rarely if ever provide fringe benefits such as health insurance. They seemed to be aware of hiring strategies across industries that are designed to avoid the payment of benefits. They spoke of ways in which their employers tried to avoid offering health insurance to them, either by only offering part-time employment or temporary work in which they were laid off before health benefits kick in. A woman in a focus group told us a common story, "No, I don't get benefits because I work 16 hours a week." Another woman, Lisa, immediately chimed in, "A lot of employers try to avoid 40 hours. They call 32–35 hours 'full-time.' But no benefits."

How, then, do our participants intend to get insurance after they leave welfare? Most were unsure. "Hopefully from my job," we heard repeatedly, but said in a tone that sounded like wishful thinking. It is doubtful that participants will be given insurance benefits in the type of jobs most of these participants are likely to find. In essence, participants were very unsure how they would get health insurance in the future after leaving welfare for work. Yet the participants were adamantly concerned about their families' health, and maintaining their ability to seek medical and dental care.

A surprising number of women did not know that they have one year in which they are entitled to the OHP, and then after that they may or may not qualify, depending on their income. Some vaguely knew that they would likely be able to purchase the OHP with premiums based on a sliding scale according to their income, but they did not know the details of eligibility, costs, or general OHP regulations. When asked about the cost of purchasing OHP, some seemed unaware that there could be costs associated with it, while others guessed on the low side, usually under $25 a month for their family. When asked about the cost of purchasing private insurance, most women seemed aware that it would likely be beyond their ability to pay for it. "It costs $350 a month. Right, like I can afford that." Another woman spoke from her experience, "At the Goodwill where I worked it was going to cost a pretty penny. There was another program that cost less, but you'd have to travel to Portland or somewhere. It's ridiculous. It's like car insurance."

It appears that families transitioning off welfare in small towns and rural communities are disadvantaged in their employment opportunities. The unemployment rate is high and wages are low. This may affect their ability to secure a job with health insurance benefits. In Oregon it is possible for most families leaving welfare to remain eligible for the OHP, given the low average wages for entry-level unskilled jobs. But we found that, despite a deep concern for health and access to health care, most focus group participants were vague when talking about plans for their future. Moreover, they were generally unaware of many OHP regulations, including that they may need to pay premiums for continued coverage.

Welfare-to-Work Program Assistance in Planning for the Future

Given the poor starting conditions of the types of jobs our participants were likely to access (i.e. part-time positions, no benefits, low wages), how do the welfare-to-work programs prepare their clients for health insurance coverage? We asked if caseworkers ever discussed health insurance options for participants that were transitioning off assistance programs. Overwhelmingly we heard that discussions of health insurance, including of the OHP, did not take place. Caseworkers did not talk about health insurance as part of the welfare-to-work program, despite its importance and its apparent interest to families transitioning off welfare. There generally was no mention of the possibility that participants may lose their insurance or that their premiums for the OHP might rise dramatically, nor any serious discussion of how to plan for future coverage. As one woman told us, in response to our question about whether these issues were brought up by caseworkers, "No, not really. Their idea is to find a job. Get a job and that's all there is to it." This was a common sentiment expressed in all seven focus groups. The participants were adamant that health insurance, and planning for insurance are not components of the welfare-to-work curriculum.

Instead, we heard that information is often transferred to clients in a haphazard manner by caseworkers, if transferred at all. One woman referred to the welfare office as a "learn by experience place," meaning that clients gained information by trial and error, and that common sources of knowledge were friends and other clients rather than caseworkers. That is, people engaged in the assistance process readily exchanged information with others about rules, regulations, options, and opportunities. This informal exchange was often viewed as more helpful than the formal exchange between caseworkers and client. This was a general phenomenon we observed across groups, sometimes explicitly stated, and other times implied. "No, they don't talk about it. I wasn't given any choice in packages. I talked to a friend."

Caseworkers are seen as gatekeepers to important services and programs. Participants suggested that caseworkers usually respond to inquiries, but that they do not always volunteer information about health insurance or other services. "All you do is read the pamphlets they give you." Caseworkers were seen to vary tremendously in their helpfulness, for example, the amount of information volunteered. "I've had ones take care of me real well. But this one doesn't even return phone calls." Another woman expressed a common sentiment when she said, "They're very different. They will tell you different things. My last one was an angel," indicating that she was more forthright in her approach. When participants realize that they can learn more about the workings of OHP from other clients and their friends than from their caseworkers, it appears that the welfare office is either incompetent or uncaring. This exacerbates the problem of determining whom to trust for information. Do you trust information from people in your same situation who have experienced the same troubles and have the same goals? Or do you trust representatives of the bureaucracy that represent access to resources while operating under incentives to reduce caseloads?

Data from these focus groups suggest that health insurance, and planning for health insurance is not a component of any welfare-to-work curriculum, or routinely part of any serious conversations. Regarding the influence of caseworkers, most comments allude to a lack of guidance from the office. To many of the participants, we were the first ones to raise these issues with them. Consequently, they had done virtually no planning for when their transitional Medicaid expires despite their serious concerns about access to health care and their previous negative experiences with being uninsured.

CONCLUSION

This research, based on data obtained from seven focus groups with 35 participants in rural communities and small towns, examined the health care needs and experiences of families transitioning off of welfare, and how they are planning for the time after their automatic benefits from the OHP expire. We found a considerable need for health care services, and heard suggestions on how the OHP could be better structured to meet their needs. Given limited employment possibilities in rural areas and small communities, the odds of obtaining employer-sponsored health insurance benefits after leaving welfare are relatively small. While some remain hopeful that their jobs will provide benefits – an unlikely prospect – others are vague and uncertain as to where benefits will come from. They hope to continue to receive the OHP, but are largely unaware of eligibility and premium requirements. Their caseworkers generally do not discuss these issues with them. Information on jobs, assistance, and health insurance was accessed informally, often in the waiting room of the welfare office. Trust of their caseworkers is often lacking, information is obtained in a haphazard fashion, and thus families transitioning off of welfare do little systematic planning for meeting the health care needs of their families. This is ironic, given the primacy that they placed upon maintaining access to the health care system.

This research provides baseline knowledge needed to address how families residing in rural areas and small towns plan for a time when their automatic benefits expire. These are critical issues because policymakers throughout the United States are struggling with ways to meet the health care needs of increasing numbers of poor families who leave the welfare system for work. Welfare reform programs rely heavily on employer-sponsored coverage after the one year of automatic Medicaid (or OHP) expires. Yet persons living outside of large cities or metropolitan areas are particularly vulnerable because of their more limited employment opportunities and the greater likelihood that employers do not offer insurance to their workers as a fringe benefit of employment. The magnitude of this problem is immense, and will only intensify over time, as the single year of transitional Medicaid is only now just beginning to expire for the initial cohorts of millions of adults and children throughout the nation.

REFERENCES

Duncan, R., Seccombe, K., & Amey, C. (1995). Changes in health insurance coverage with rural-urban environments. *Journal of Rural Health, 11*, 169–177.

Families USA (1999). *Losing health insurance: The unintended consequences of welfare reform.*

Families USA (2001). *Key facts about the uninsured.* Retrieved from available online: www.familiesusa.org

Garrett, B., & Holahan, J. (2000, Jan–Feb). Health coverage after welfare. *Health affairs*, available online: www.projhope.org/HA/janfeb001/190116.htm

Greenberg, M. (1998). *Participation in welfare and medicaid enrollment* (Technical. Report No. Issue Paper). Washington, DC: Henry J. Kaiser Foundation.

Heymann, S., & Earle, A. (1999). The impact of welfare reform on parents' ability to care for their children's health. *American Journal of Public Health, 89*, 502–505.

Kaiser Commission on Medicaid and the Uninsured (1998). *The uninsured and their access to health care.* Washington, DC: Henry J. Kaiser Family Foundation.

Kaiser Commission on Medicaid and the Uninsured (2002, July). *Uninsured in America: Is health insurance adequate?* Washington, DC: Henry J. Kaiser Family Foundation.

Klein, R. (2002). *Children losing health insurance.* Washington, DC: Families USA.

Larkin, H. (1999). *Are more uninsured an unintended consequence of welfare reform?* (Advances No. Issue 2). Princeton, NJ: Robert Wood Johnson Foundation.

Loprest, P. (2001, April). *How are families that left welfare doing? A comparison of early and recent welfare leavers* (Tech. Rep. No. Series B, No. B-36). Washington, DC: Urban Institute Press.

Loprest, P. (2002). Making the transition from welfare to work: Successes but continuing concerns. In: A. Weil & K. Feingold (Eds), *Welfare Reform: The Next Act* (pp. 17–31). Washington, DC: Urban Institute Press.

Loprest, P., & Acs, G. (1996). *Profile of disability among families on AFDC.* Washington, DC: Urban Institute Report to the Henry J. Kaiser Family Foundation.

Morgan, D. L. (1997). *Focus groups as qualitative research*. Thousand Oaks, CA: Sage.

Murray, C. (1984). *Losing ground: American social policy, 1950–1980*. New York, NY: Basic Books.

Murray, C. (1988). *In pursuit of happiness and good government*. New York, NY: Simon and Schuster.

O'Brien, E., & Feder, J. (1998). *How well does the employment-based health insurance system work for low-income families?* (Issue Paper). Washington, DC: Henry J. Kaiser Foundation.

Oregon Health Plan Policy Research (2003). *2003 Summary Tables*. Retrieved from available online: www.ohppr.state.or.us/data/ops/data_ops_index.htm

Rowe, G. (2000). *Welfare rules databook: State TANF policies as of July 1999*. Washington, DC: Urban Institute.

Seccombe, K. (1995). Health insurance and use of services among low income elders: Does residence influence the relationship? *Journal of Rural Health*, *11*, 86–97.

Seccombe, K. (1999). *So you think I drive a cadillac? Welfare recipient's perspectives on the system and its reform*. Needham Heights: Allyn & Bacon.

U.S. Department of Health and Human Services (2002). *Statistics: U.S. Welfare caseload information*. Retrieved from Available online at: www.acf.dhhs.gov/news/stats/newstat2.shtml

Weigers, M., Weinick, R., & Cohen, J. (1998). Children's health insurance, access to care, and health status: New findings. *Health Affairs*, *17*, 127–136.